AF893548

PACKAGE DESIGN IN JAPAN VOL.5

PACKAGE DESIGN IN JAPAN VOL.5

EDITED BY JAPAN PACKAGE DESIGN ASSOCIATION AND RIKUYO-SHA

PACKAGE DESIGN IN JAPAN VOL. 5

Editorial Director	Japan Package Design Association
Book Designer	Tetsuya Ota
Photograher	Ken Abe Bank Publicity Co., Ltd. (Special Feature)
Translator	Chine Hayashi
Editorial Staff	Rikuyo-sha Izumi Hosokawa/Masami Hirose
Publisher	Hashimoto Shuhei Rikuyo-sha Publishing, Inc. Shizuoka Bank Bldg., 19-12, 2-chome, Shinjuku, Shinjuku-ku, Tokyo, Japan Phone: 03-3354-4020 Fax: 03-3352-3106

© 1993 Rikuyo-sha Publishing, Inc.
Printed and Bound in Japan

着眼と冒険と社会性

日本パッケージデザイン協会（JPDA）が主催するこの「日本パッケージデザイン大賞」は、デザインを見地とするコンペティションとして定着をみてきました。この賞が担っている責任の重みをいっそう感じています。それゆえ運営の改善にもつねづね務め、今回、賞の体系を新たにしました。

パッケージが現代社会の経済と生活にとって不可欠であることは今さらいうまでもありません。そのためパッケージは、つねに変化してゆく時代の要請と課題に沿いながら、市場、生活者、社会にとって適切な用と美を提案するものでなければなりません。

今回の応募総数は933点で、前回の1,220点より下廻りましたが、それは、低迷する経済状況あるいは重要性を増している環境問題等が、より本格的なデザインを求めているからでしょう、応募する段階ですでに厳選されたことによると思えます。それゆえ今回は激戦になりました。

1次審査で354点が入選し、その中から2次審査で63点が入賞候補となり、審査員の率直で厳しい論議を経て23点が入賞しました。大賞・金賞・銀賞・特別賞のうち、特別賞は銅賞ではなく、特に評価したい特質によって選定したものです。

審査を通じて感じたのは、全体にデザイン完成度が高まり、この点だけでは優劣をつけ難かったことです。そのため着眼の新しさ、造形の冒険、エコロジー対応など社会性の有無が問われました。エコロジー問題に関しては、ともすればありがちな紋切り型あるいは狭い視野での観念的主張に引きずられないようにと、この問題に通じた包装技術家を特別審査員に迎えました。本書はその小論を載せてありますのでご参考としてください。

真剣かつ公平な審査によって入賞・入選した全作品を収録してある本書は時代の映写幕でもあります。そこから次の時代への動きをお読みとりいただけるならば幸いです。

金子修也
社団法人日本パッケージデザイン協会理事長

Conception, Challenging Forms and Social Consideration

The Japan Package Design Award sponsored by the Japan Package Design Association is one of the established design competitions in Japan. We have been trying to improve the administration of the competition, and as a step towards this, we have changed the awarding system, with a revised award system being implemented beginning with the Japan Package Design Award '93 competition.

Packages are indispensable in modern life. Package design should provide both the market and consumers with functionality and aesthetic beauty.

The competition this time saw 933 entries, less than the figure of the previous occasion which was 1,220. It is a reflection of the lingering recession and increasingly crucial environmental concerns that demand genuinely high quality packages. It is likely that serious self-scrutiny was made at the stage of application, hence, more fierce competition.

Among the entries, 354 works were screened, 63 works of which were further shortlisted, and after candid and strict selection, 23 works won the Awards; the Grand Prix, Golden Award, Silver Award, and Special Award. The Special Awards were given to those packages with outstanding merits.

The general design standard of the entries was high, and it was difficult to evaluate them on the basis of design quality alone. Therefore, other criteria such as, novelty, challenging attempts in form and, social considerations such as ecological devices were adopted. With regard to ecological considerations, in particular, Mr. Osamu Ishida, an expert in package manufacturing was invited to be a judge, so as to avoid stereotyped, and too narrowly conceptual discussions. His essay is included in this book for your reference.

This book containing all the award winning and other selected works is a screen onto which the present time is projected. It is our hope that the readers will be able to clearly see this time period from the pages of this book.

Shuya Kaneko
Executive Director
Japan Package Design Association

審査員
Awarding Judges

審査委員長 *Chief Awarding Judge*	金子修也	*Shuya Kaneko*
審査員 *Awarding Judges*	青木茂吉	*Shigeyoshi Aoki*
	池田 毅	*Takeshi Ikeda*
	犬塚達美	*Tatsumi Inuzuka*
	薄井寿一	*Toshiichi Usui*
	大森重志	*Shigeshi Omori*
	岡沢 豪	*Takeshi Okazawa*
	岡田宏三	*Kozo Okada*
	奥村昭夫	*Akio Okumura*
	鹿目尚志	*Takashi Kanome*
	川路ヨウセイ	*Yosei Kawaji*
	桑 和美	*Kazumi Kuwa*
	佐竹愼一郎	*Shin'ichiro Satake*
	高橋 敏	*Ving Takahashi*
	中山英子	*Eiko Nakayama*
	中山紀子	*Michiko Nakayama*
	西村英明	*Hideaki Nishimura*
	広橋桂子	*Keiko Hirohashi*
	松橋紘一	*Koichi Matsuhashi*
	村 禎介	*Teisuke Mura*
	和田 亨	*Toru Wada*

Contents

略記号　Abbreviation

PL —— Planner
CD —— Creative Director
CAD — Chief Art Director
AD —— Art Director
D —— Designer
Adv —— Adviser
BD —— Bottle Designer
GD —— Graphic Designer
PD —— Product Designer
Const — Construction Designer
LO —— Logotype Designer
LD —— Label Designer
CA —— Calligrapher
IL —— Illustrator
Art —— Artist
P —— Photographer
Copy — Copywriter
AC —— Art Coordinator
ST —— Stylist
AG —— Agent
PR —— Producer
CL —— Client

入賞作品

Winning Works

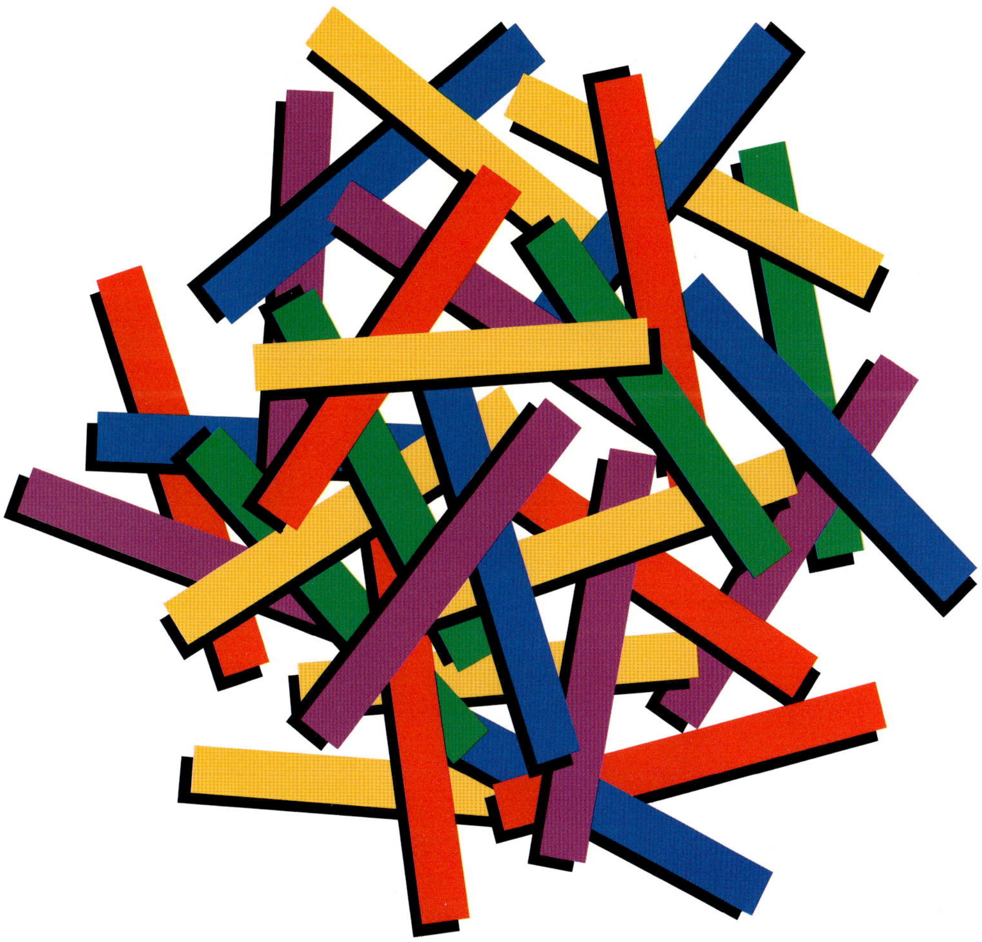

大賞 | Grand Prix

1 フェミニテ・デュ・ボワ
AD 杉浦俊作
D 塩川 稔
杉浦俊作
Adv セルジュ・ルタンス
CL 資生堂

Fragrance
AD Shunsaku Sugiura
D Minoru Shiokawa
Shunsaku Sugiura
Adv Serge Lutens
CL Shiseido

1

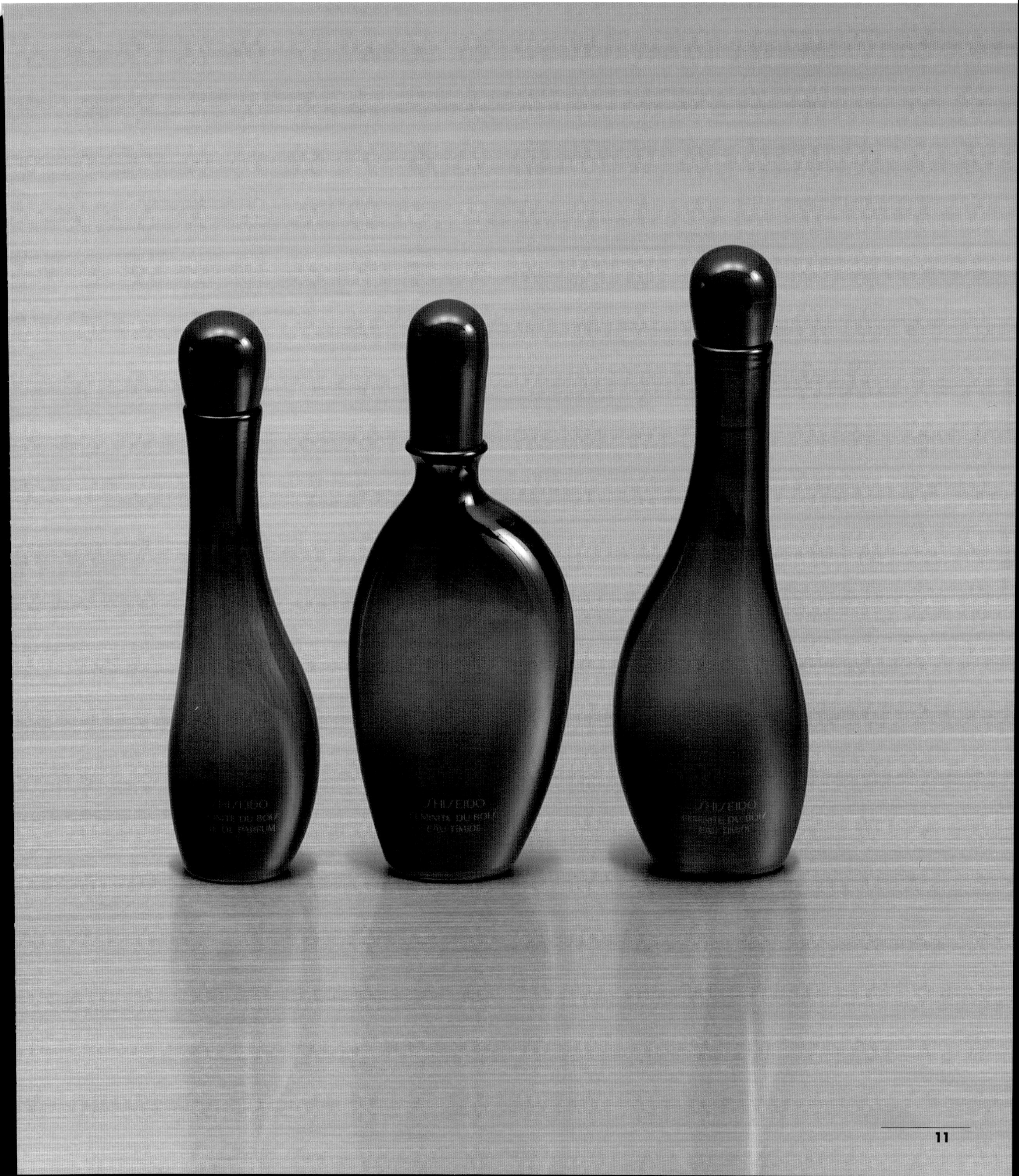

金賞 | Golden Award

2 モールド パコ

AD 鹿目尚志
D 鹿目尚志
Const 鹿目尚志
CL 包む

Wine Gift Box

AD Takashi Kanome
D Takashi Kanome
Const Takashi Kanome
CL Tsutsumu

2

金賞 Golden Award

3 セルフコンシャス シリーズ

CD 中山紀子
AD 塙冨士雄
D 篠 真澄
鏑木ユキ江
上東吉明
ヴァレリー・ベルナール
CL コーセー

Cosmetics

CD Michiko Nakayama
AD Fujio Hanawa
D Masumi Shino
Yukie Kaburaki
Yoshiaki Uehigashi
Valerie Bernard
CL Kosé

3

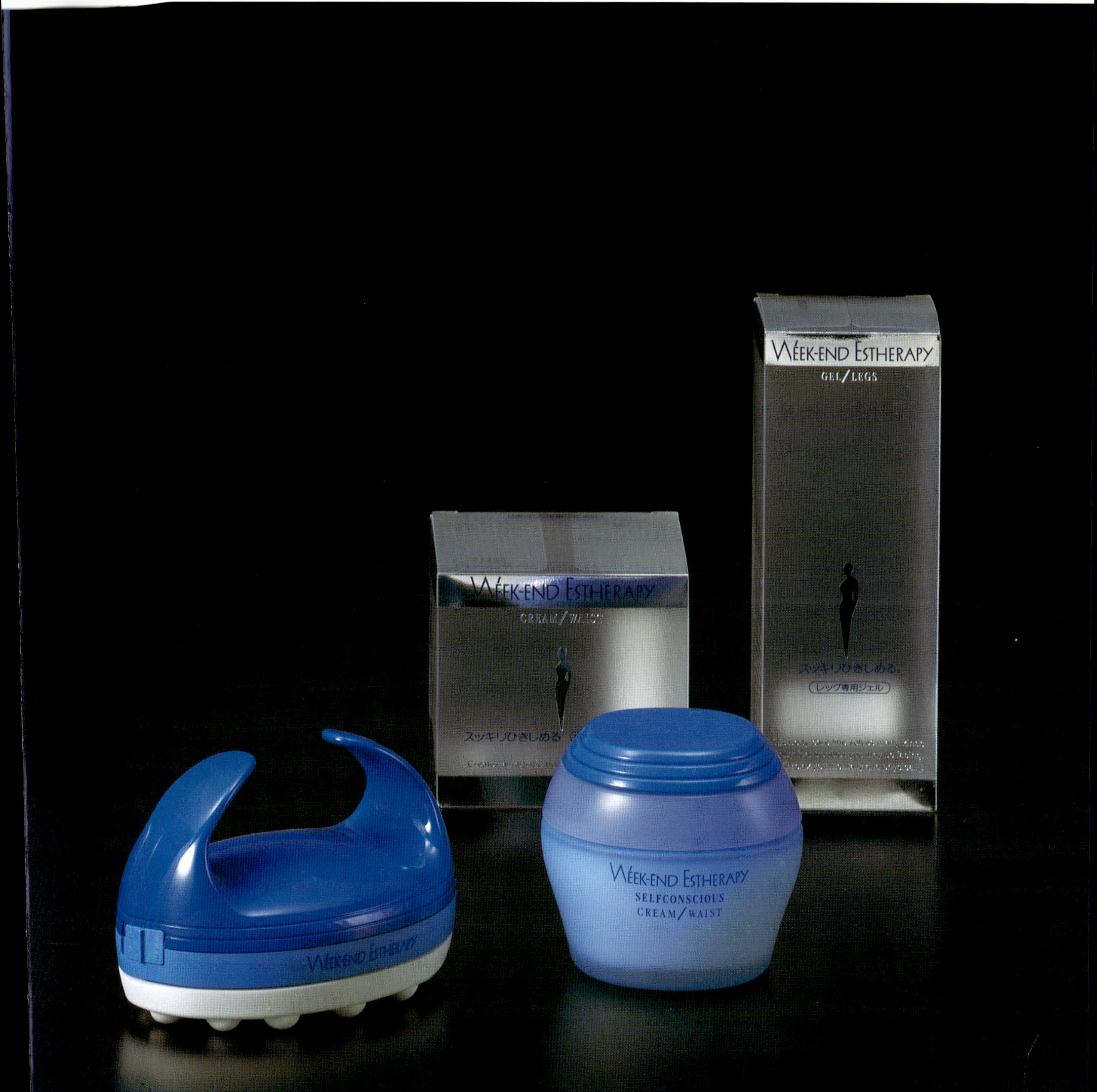

金賞 Golden Award

4 ポーラ ヴァカンシィ
AD 大野淳一郎
D 有田 賢
CL ポーラ化粧品本舗

Cosmetics
AD Jun'ichiro Ono
D Satoshi Arita
CL Pola

4

金賞 | Golden Award

5 キッコーマン卓上しょうゆびん100ml
AD GKグラフィックス
D GKグラフィックス
CL キッコーマン

Soy Sauce
AD GK Graphics
D GK Graphics
CL Kikkoman

5

銀賞 Silver Award

6 タリアセンコーヒー
AD GKグラフィックス
D GKグラフィックス
CL タリアセン

Coffee
AD GK Graphics
D GK Graphics
CL Taliesin

6

銀賞 Silver Award

7 UNO

AD 松永 真
山下正和
足立和彦
D 松永 真
鳩貝浩司
酒井重彌
CL 資生堂

Men's Cosmetics

AD Shin Matsunaga
Masakazu Yamashita
Kazuhiko Adachi
D Shin Matsunaga
Koji Hatogai
Sigeya Sakai
CL Shiseido

7

銀賞 | Silver Award

8 ポーラ システムクロノス
AD 片田博史
D 三輪万寿夫
長瀬正英
CL ポーラ化粧品本舗

Cosmetics
AD Hirofumi Katada
D Masuo Miwa
Masahide Nagase
CL Pola

銀賞 Silver Award

9 NAN-PAO

CD 鞍井修一
原 恒男
AD 奥村昭夫
D 楠本 毅
AG 電通関西支社
CL 田辺製薬

Medicine

CD Shuichi Kurai
Tsuneo Hara
AD Akio Okumura
D Takeshi Kusumoto
AG Dentsu, Kansai
CL Tanabe Seiyaku

銀賞 | Silver Award

10 宝焼酎 スーパー純
AD 佐藤 卓
D 佐藤 卓
PL ユニゾンネットワーク
CL 宝酒造

Spirits
AD Taku Sato
D Taku Sato
PL Unison Network
CL Takara Shuzo

10

銀賞 | Silver Award

11 サントリーワイン頒布会'91秋
日本の伝統美コース

AD 牛島志津子
D 牛島志津子
今出栄美子
CL サントリー

Wine

AD Shizuko Ushijima
D Shizuko Ushijima
Emiko Imade
CL Suntory

銀賞 | Silver Award

12 サントリースキットボトル
リザーブ　オールド　V.S.O.P

CD 横尾 浩
AD 藤本勝之
D 藤本勝之
寿精版印刷デザイン室
IL アイ工房
CL サントリー

Whisky Brandy

CD Hiroshi Yokoo
AD Katsuyuki Fujimoto
D Katsuyuki Fujimoto
Kotobuki Seihan Printing, Design Room
IL Ai Kobo
CL Suntory

12

銀賞 | Silver Award

13 サントリービア・ヌーボー **1991年夏**
AD 大門敏彦
D ケン・ドーン
CL サントリー

Beer
AD Toshihiko Daimon
D Ken Done
CL Suntory

13

銀賞 Silver Award

14 ダンガリー
AD 鹿目尚志
D 鹿目尚志
BD 鹿目尚志
CL 資生堂

Men's Cosmetics
AD Takashi Kanome
D Takashi Kanome
BD Takashi Kanome
CL Shiseido

14

銀賞 Silver Award

15 ダスキン洗濯洗剤シリーズ

AD 犬塚達美
D 犬塚達美
佐藤勝則
BD 犬塚達美
近藤純夫
P 安孫子貴嗣
CL ダスキン

Detergent

AD Tatsumi Inuzuka
D Tatsumi Inuzuka
Katsunori Sato
BD Tatsumi Inuzuka
Sumio Kondo
P Takatsugu Abiko
CL Duskin

15

特別賞 Special Award

色、形、文字のもたらす躍動感と楽しさ
Dynamism and beauty brought about by colors, forms and lettering

16 資生堂パーラー バレンタインチョコレート
AD 仲條正義
D 仲條正義
CL 資生堂パーラー

Chocolates
AD Masayoshi Nakajo
D Masayoshi Nakajo
CL Shiseido Parlour

16

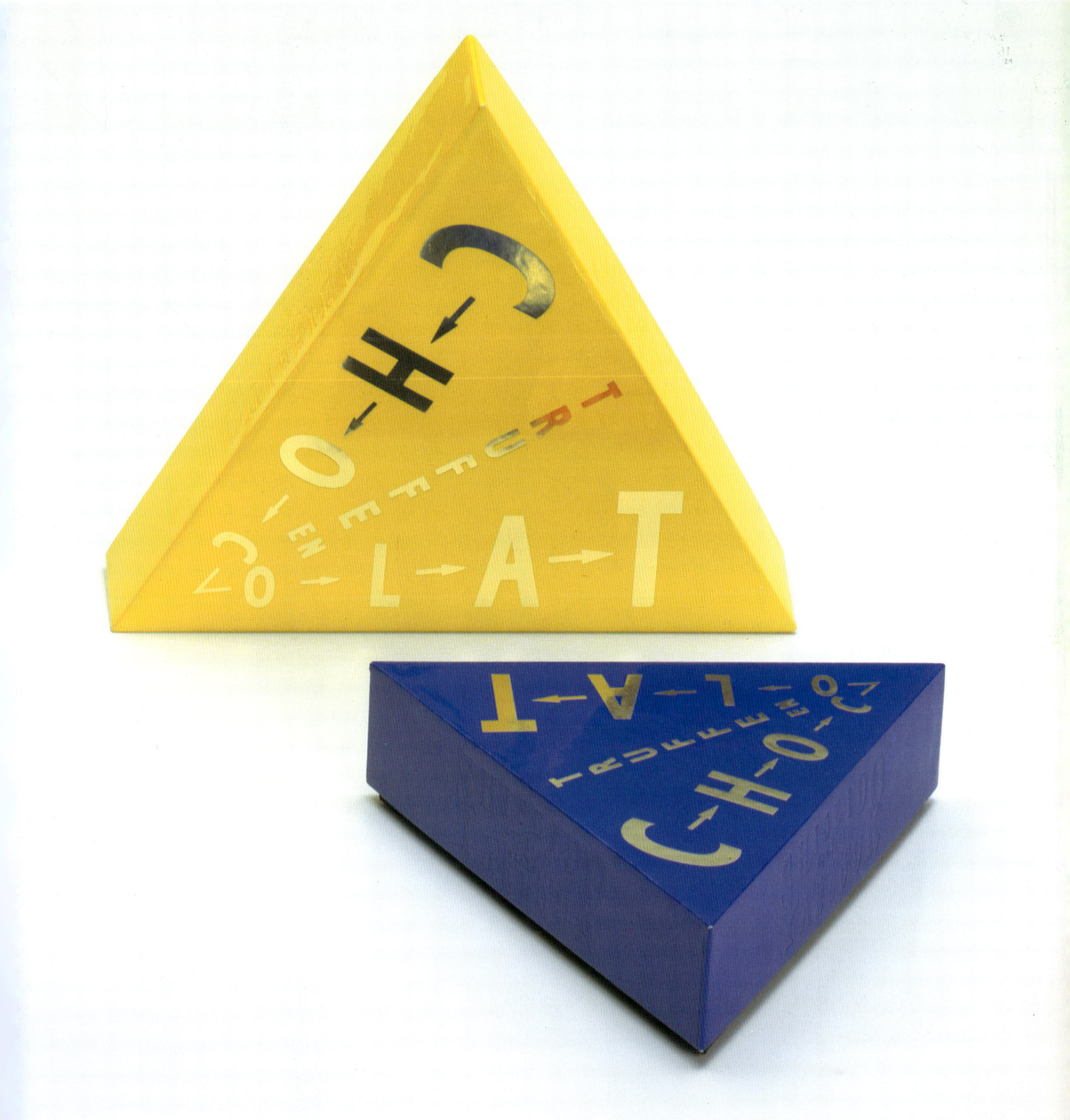

特別賞 Special Award

片手で使える便利さとキッチンを明るくするデザイン
By design and convenience that can be used by one hand

17 ハウス デイズコショー
CD 小笠原輝彦
AD 佐藤 卓
D 佐藤 卓
CL ハウス食品工業

Spice
CD Teruhiko Ogasawara
AD Taku Sato
D Taku Sato
CL House Food Industrial

17

特別賞 Special Award

中南米の香りをファッショナブルに表現
Fashionable expression of Latin American flavor

18	ゴストーコーヒー トポス	**Coffee**	
PL	山崎勝之	PL	Katsuyuki Yamazaki
AD	久保 繁	AD	Shigeru Kubo
D	久保 繁	D	Shigeru Kubo
Const	高橋寿介	Const	Toshisuke Takahashi
CL	アイ・シー電子工業	CL	I C Electronics Industry

特別賞 Special Award

白地に赤の独特のロゴタイプ
Unique red logo on the white background

19 Dolight

AD 廣 哲夫
D 村 禎介
鈴木 良彦
比留間英二
CL 資生堂

Drink Food

AD Tetsuo Hiro
D Teisuke Mura
Yoshihiko Suzuki
Eiji Hiruma
CL Shiseido

19

特別賞 | Special Award

伝統的で新しい日本酒の表現
Traditional yet fresh expression of Japanese Sake

20 秋鹿純米大吟醸一貫造り
AD 坂東由章
D 坂東由章
LO 荻野丹雪
CL 秋鹿酒造

Sake
AD Yoshiaki Bando
D Yoshiaki Bando
LO Tansetsu Ogino
CL Akishika Sake Brewing

20

特別賞 | Special Award

紙と同じに燃やせる素材を使った、過剰包装廃止の提案

Initiative for abolishing excessive packaging through using a burnable material such as paper

21 BAG・AGE

PL シャープ通信オーディオデザインセンター
D シャープ通信オーディオデザインセンター
AG 厚木プラスチック
CL シャープ

Cordlessphone

PL Sharp Communication & Audio Systems Design Center
D Sharp Communication & Audio Systems Design Center
AG Atsugi Plastic
CL Sharp

特別賞 Special Award

22 カネボウ化粧品 **ECシリーズ**

PL 林 洋二
AD 池田 毅
D 延安敬子
CL カネボウ化粧品

Cosmetics

PL Yoji Hayashi
AD Takeshi Ikeda
D Takako Nobuyasu
CL Kanebo Cosmetics

容器の樹脂半減と再生紙使用のエコロジカル・コスメティックス
Ecologically-friendly cosmetic package using recycled paper and reducing the use of resin by half

22

特別賞 Special Award

廃棄乾電池の環境問題とその対策を表現
Expression of the company's environmental concern by collecting used batteries

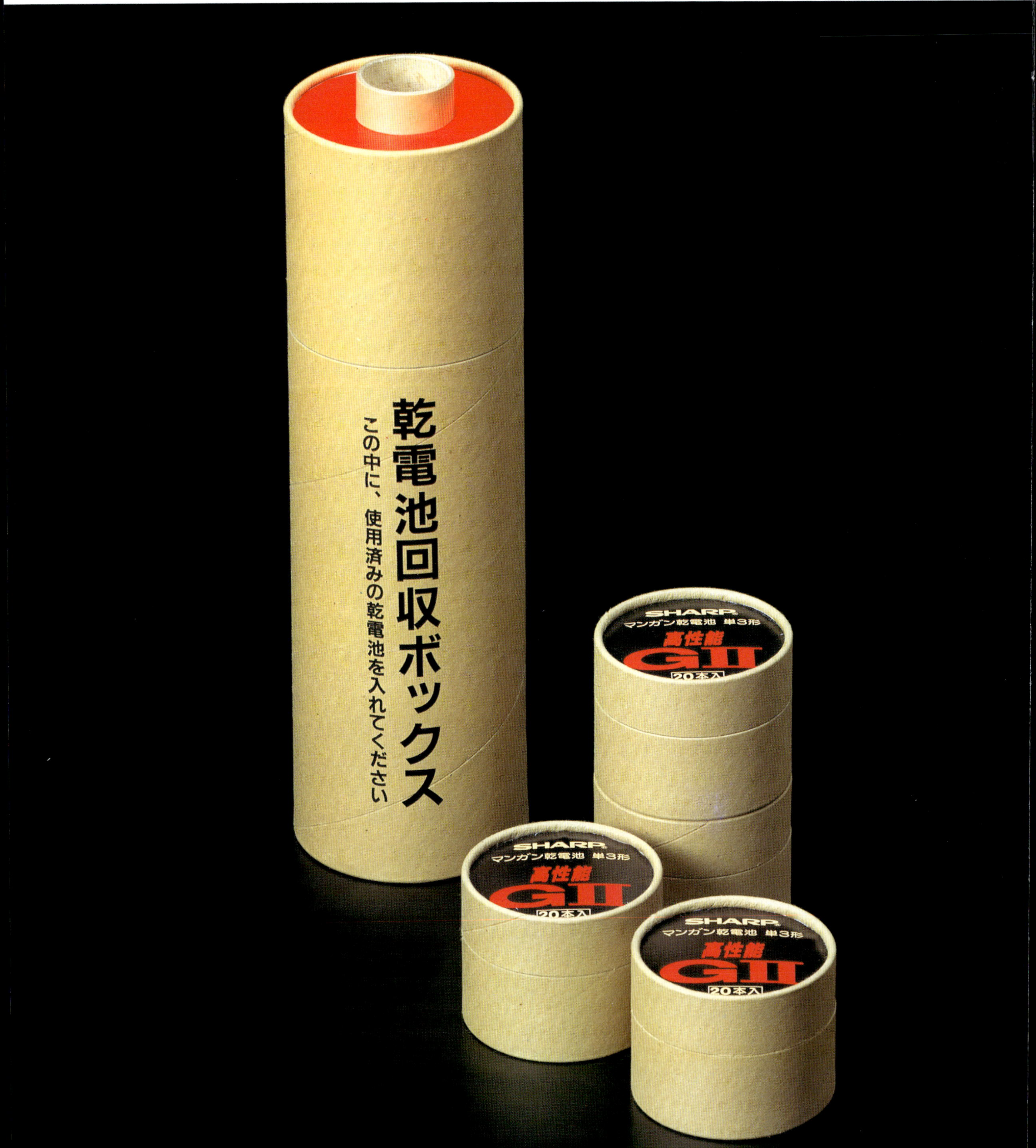

23 乾電池の回収／販促パッケージ
PL シャープ通信オーディオデザインセンター
D シャープ通信オーディオデザインセンター
CL シャープ

Battery
PL Sharp Communication & Audio Systems Design Center
D Sharp Communication & Audio Systems Design Center
CL Sharp

食品 菓子

Food Stuff/Confectionery

24

24 シスコーンシリーズ
AD 木全時彦
塩津博基
D 半田広喜
川崎裕司
CL 日清シスコ

Cereal
AD Tokihiko Kimata
Hiroki Shiozu
D Hiroki Handa
Hiroshi Kawasaki
CL Nissin Cisco

25 カルビー ブラウンシュガーシリーズ

PL 電通
AD 佐藤 卓
D 佐藤 卓
西塔美由紀
CL カルビー

Cereal

PL Dentsu
AD Taku Sato
D Taku Sato
Miyuki Saito
CL Calbee Foods

25

26 フォーモスト ブルー シール
アイスクリーム パーティーポップ

AD 佐藤 卓
D 佐藤 卓
CL フォーモスト ブルー シール

Ice Cream

AD Taku Sato
D Taku Sato
CL Foremost Blue Seal

26

27

27 ジャネフ ノンシュガージャム

AD 山下勇三
D 伊藤 優
IL 山下勇三
CL キユーピー

Jam

AD Yuzo Yamashita
D Masaru Ito
IL Yuzo Yamashita
CL Q.P.

28

28 バラッティ ジャム

PL 佐藤洋公
AD アルファキュービック
D 久保田マリア
LO 久保田マリア
IL 久保田マリア
CL アルファキュービック・エンタープライズ

Jam

PL Hirooki Sato
AD Alpha Cubic
D Maria Kubota
LO Maria Kubota
IL Maria Kubota
CL Alpha Cubic Enterprize

29

29 バラッティ 蜂蜜

PL 佐藤洋公
AD アルファキュービック
D 久保田マリア
LO 久保田マリア
IL 久保田マリア
CL アルファキュービック・エンタープライズ

Honey

PL Hirooki Sato
AD Alpha Cubic
D Maria Kubota
LO Maria Kubota
IL Maria Kubota
CL Alpha Cubic Enterprize

30

30 アルトレコーゼ スパイスシリーズ

PL 佐藤洋公
AD アルファキュービック
D 久保田マリア
LO 久保田マリア
IL 久保田マリア
CL アルファキュービック・エンタープライズ

Spice

PL Hirooki Sato
AD Alpha Cubic
D Maria Kubota
LO Maria Kubota
IL Maria Kubota
CL Alpha Cubic Enterprize

31 アルトレコーゼ ソーセージ

PL 佐藤洋公
AD アルファキュービック
D 久保田マリア
LO 久保田マリア
CL アルファキュービック・エンタープライズ

Sausage

PL Hirooki Sato
AD Alpha Cubic
D Maria Kubota
LO Maria Kubota
CL Alpha Cubic Enterprize

31

32 アルトレコーゼ ハーブ入りオリーブオイル

PL 佐藤洋公
AD アルファキュービック
D 久保田マリア
LO 久保田マリア
IL 久保田マリア
CL アルファキュービック・エンタープライズ

Olive Oil

PL Hirooki Sato
AD Alpha Cubic
D Maria Kubota
LO Maria Kubota
IL Maria Kubota
CL Alpha Cubic Enterprize

32

33

33 ハウスクッキングスパイス

CD 池田真理
AD 佐藤 卓
D 佐藤 卓
西塔美由紀
CL ハウス食品工業

Spice

CD Mari Ikeda
AD Taku Sato
D Taku Sato
Miyuki Saito
CL House Food Industrial

34

34 マルキン
しぼりたて 生しょうゆ

AD 小宮 茂
BD 小宮 茂
小宮麻美子
GD 小宮 茂
小宮 麻美子
LO 足立花鏡
CL 丸金醤油

Soy Sauce

AD Shigeru Komiya
BD Shigeru Komiya
Mamiko Komiya
GD Shigeru Komiya
Mamiko Komiya
LO Kakyo Adachi
CL Marukin Shoyu

35 正田醤油 ムラサキゴノミ

AD 正田 隆
D 川路ヨウセイ
小澤満佐子
IL 片山智恵子
CL 正田醤油

Say Sauce

AD Takashi Shoda
D Yosei Kawaji
Masako Ozawa
IL Chieko Katayama
CL Shoda Shoyu

35

36

36 キユーピー
1/3 ドレッシング

AD 山下勇三
D 鈴木浩子
BD 山下勇三
IL 須賀朋樹
CL キユーピー

Dressing

AD Yuzo Yamashita
D Hiroko Suzuki
BD Yuzo Yamashita
IL Tomoki Suga
CL Q.P.

37

37 ミツカン
オイル&ビネガーシリーズ

CD 堂本 哉
AD 和久井義男
D 和久井義男
P 川合 泉
CL 中埜酢店

Dressing

CD Hajime Domoto
AD Yoshio Wakui
D Yoshio Wakui
P Izumi Kawai
CL Nakano Vinegar

38

38 キユーピードレッシング

AD 山下勇三
D 須賀朋樹
BD 山下勇三
IL 山下勇三
CL キユーピー

Dressing

AD Yuzo Yamashita
D Tomoki Suga
BD Yuzo Yamashita
IL Yuzo Yamashita
CL Q.P.

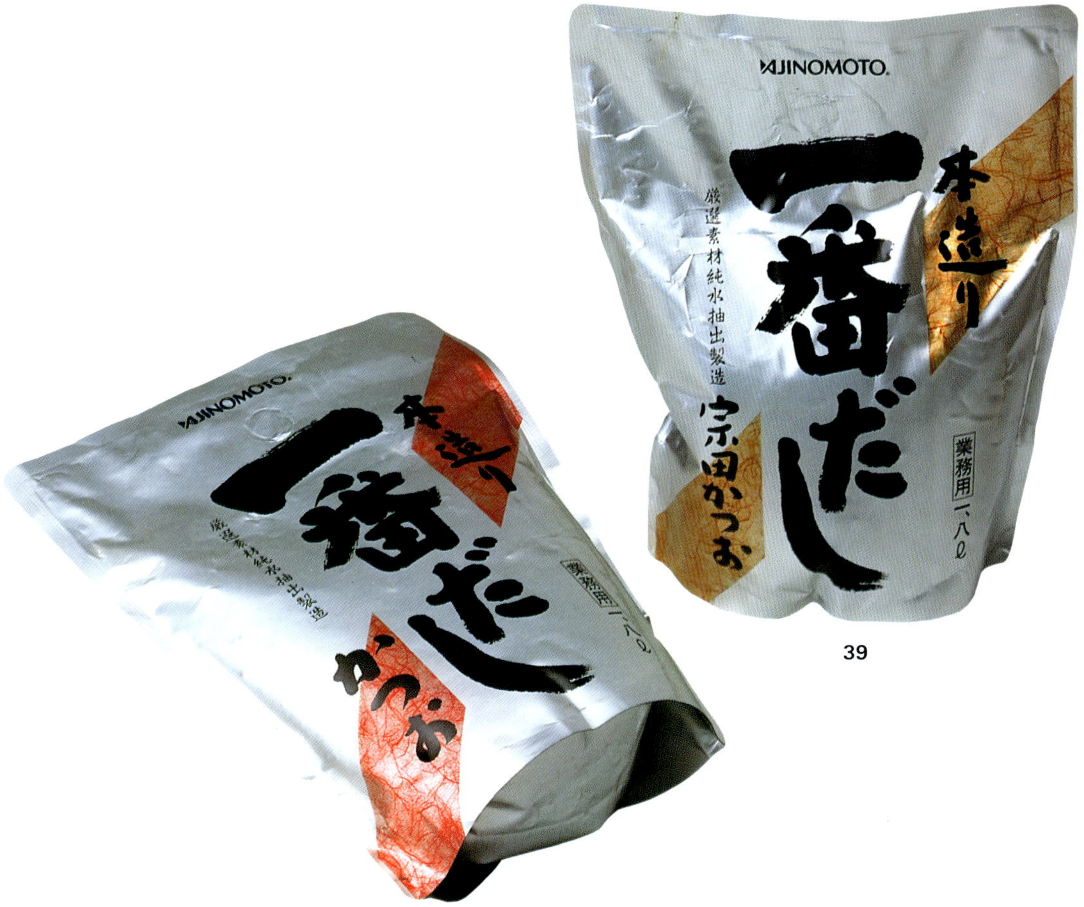

39

39 一番だし
CD 梅田悦史
AD 島崎紘而
D 上田博司
平田みか子
CL 味の素

Seasoning
CD Etsufumi Umeda
AD Koji Shimazaki
D Hiroshi Ueda
Mikako Hirata
CL Ajinomoto

40 ほんだし／
ほんだしかつお・こんぶだし
CD 梅田悦史
AD 宮地睦明
島崎紘而
D 入沢滋彦
上田博司
CL 味の素

Seasoning
CD Etsufumi Umeda
AD Mutsuaki Miyaji
Koji Shimazaki
D Shigehiko Irisawa
Hiroshi Ueda
CL Ajinomoto

40

41 豆腐 自然屋

AD シヴァデザイン事務所
D 清水直明
CL 自然屋

Tofu

AD Siva Design Office
D Naoaki Shimizu
CL Jinenya

42 日清製粉 風味一番特選つゆ
関東風 関西風

AD 竹内 豊
D 岡田宏三
田中春雄
山下隆夫
CL 日清製粉

Soup

AD Yutaka Takeuchi
D Kozo Okada
Haruo Tanaka
Takao Yamashita
CL Nissin Flour Milling

41

42

43 季の奏 豆腐セット

AD 松岡正巳
D 松岡正巳
ふじいのぶこ
LO 南 明子
CL プラスパフーズ味道楽

Tofu

AD Masami Matsuoka
D Masami Matsuoka
Nobuko Fujii
LO Akiko Minami
CL Prosper Foods Ajidoraku

43

44

44 レンジでチンするカレーうどん

CD 山口泰民
AD 岡田宏三
D 高橋宏明
Copy 田村嘉野
CL 雪印乳業

Frozen Food

CD Yasutami Yamaguchi
AD Kozo Okada
D Hiroaki Takahashi
Copy Yoshino Tamura
CL Snow Brand Milk Products

45

45 カンタブランチ

AD 白井雄司
田坂厚生
D 川路ヨウセイ
小澤満佐子
CL ハウス食品工業

Spagetti

AD Yuji Shirai
Atsuo Tasaka
D Yosei Kawaji
Masako Ozawa
CL House Food Industrial

46 おかゆシリーズ
あさがゆ／おかゆすなっく

PL 堂本食品
浜本京子
AD 桑 和美
D 桑 和美
P 浅野 勇
CL 堂本食品

Rice Gruel

PL Domoto Food
Kyoko Hamamoto
AD Kazumi Kuwa
D Kazumi Kuwa
P Isamu Asano
CL Domoto Food

46

47 アヲハタのおかゆ

AD 山下勇三
D 伊藤 優
P 松井茂信
Copy 松本恵子
CL キユーピー

Rice Gruel

AD Yuzo Yamashita
D Masaru Ito
P Shigenobu Matsui
Copy Eiko Matsumoto
CL Q.P.

47

48 赤ちゃん村

AD 村 武史
D 秋吉喜久子
鈴木良彦
CL 明治乳業

Baby Food

AD Takeshi Mura
D Kikuko Akiyoshi
Yoshihiko Suzuki
CL Meiji Milk Products

48

49

49 咖喱工房
AD 白井雄司
田坂厚生
D 川路ヨウセイ
近藤和磨
CL ハウス食品工業

Curry
AD Yuji Shirai
Atsuo Tasaka
D Yosei Kawaji
Kazuma Kondo
CL House Food Industrial

50 いろいろ野菜スープ
AD 坪谷一利
D 坪谷一利
CL 明治製菓

Soup
AD Kazutoshi Tsuboya
D Kazutoshi Tsuboya
CL Meiji Seika Kaisha

50

51 カリー元年
CD 蟻田善造
AD 中川憲造
D 延山博保
森上 暁
川畑直道
CL ロック・フィールド

Curry
CD Zenzo Arita
AD Kenzo Nakagawa
D Hiroyasu Nobuyama
Satoshi Morikami
Naomichi Kawahata
CL Rock Field

51

52

52 三幸梅パッケージシリーズ
PL ロクデザインサービス
AD 武友正修
D 武友朋子
Copy 武友正修
LO 武友朋子
村田東嶽
CL 三幸農園

Pickled Plum
PL Roku Design Service
AD Masanobu Taketomo
D Tomoko Taketomo
Copy Masanobu Taketomo
LO Tomoko Taketomo
Togoku Murata
CL Sanko Nouen

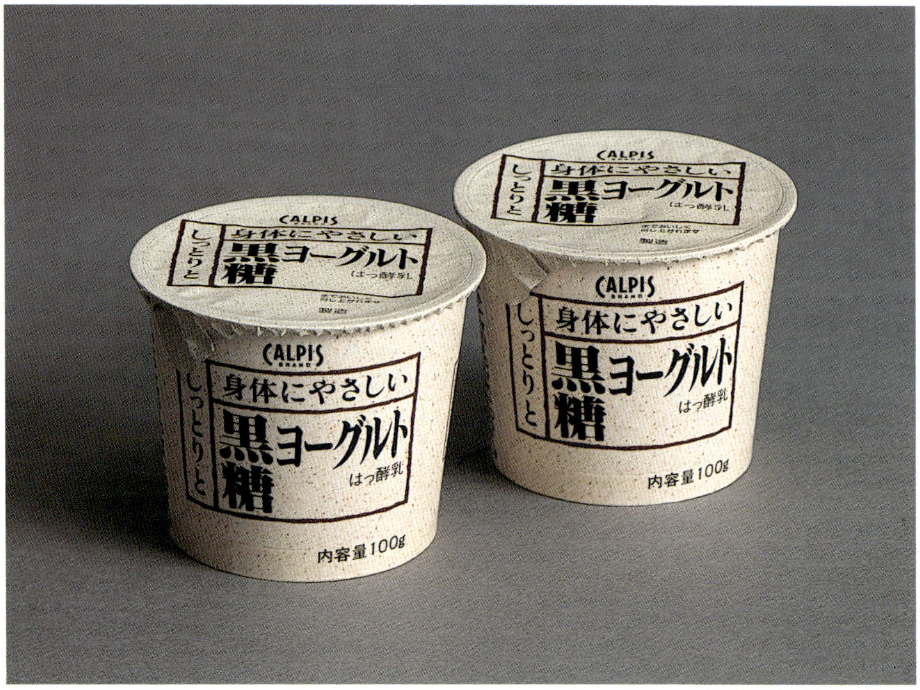

53

53 カルビス黒糖ヨーグルト
CD 和田 亨
D 中田淑子
CL カルビスフーズサービス

Yoghurt
CD Toru Wada
D Yoshiko Nakada
CL Calpis Food Service

55

55 ホウライ乳製品シリーズ
CD 麹谷 宏
AD 入江健介
D 入江健介
CL ホウライ

Dairy Products
CD Hiroshi Kojitani
AD Kensuke Irie
D Kensuke Irie
CL Horai

54 紅花ハイオレインマーガリン
AD 石川 源
D 秋山香代子
IL 藤本賢二
CL 紅花食品

Margarine
AD Minato Ishikawa
D Kayoko Akiyama
IL Kenji Fujimoto
CL Benibana Foods

54

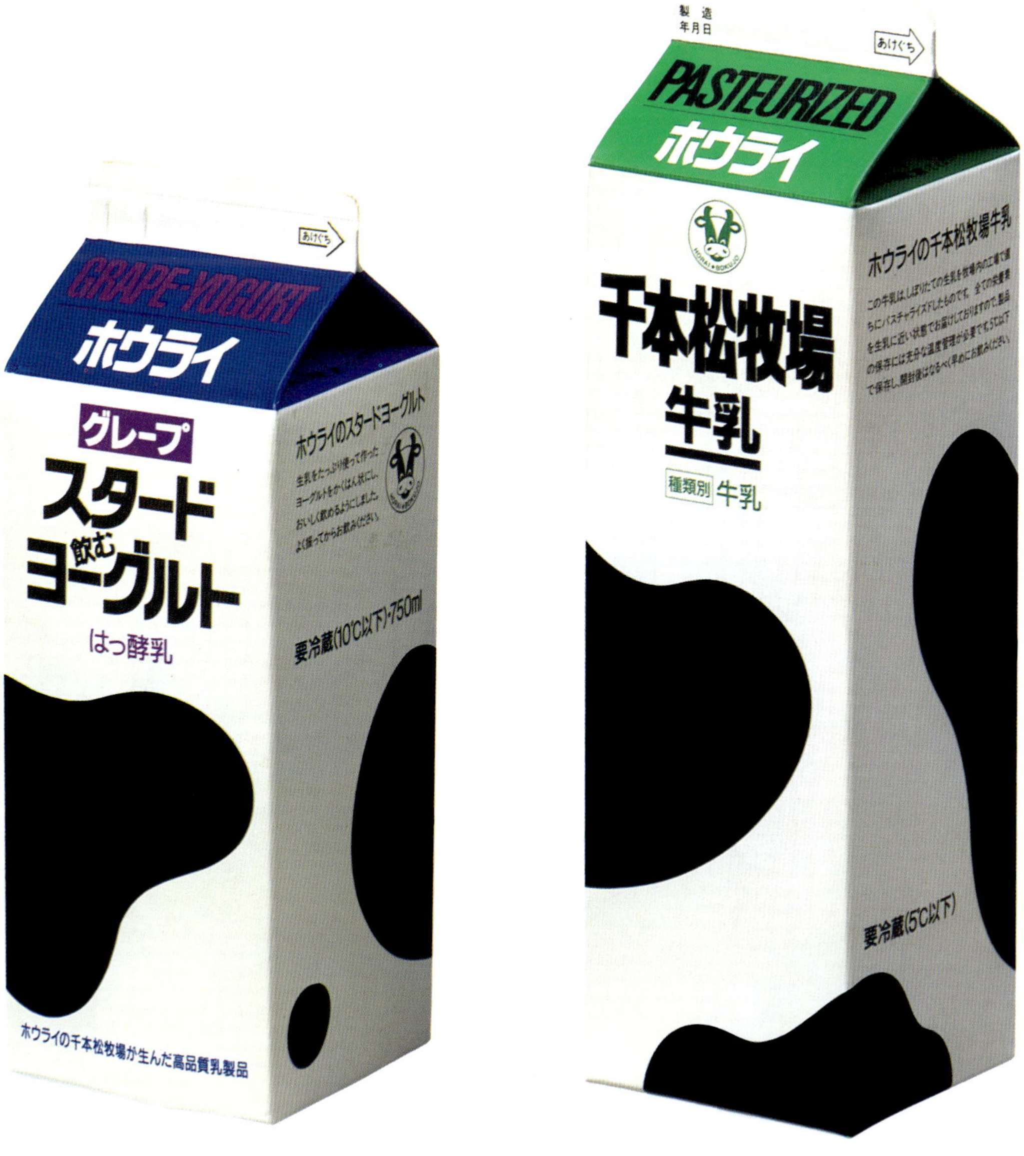

56 セシル

CD 中田一弘
AD 駒 芳昭
　　金澤一興
D 金澤一興
CL 江崎グリコ

Chocolate

CD Kazuhiro Nakada
AD Yoshiaki Koma
　　Kazuoki Kanazawa
D Kazuoki Kanazawa
CL Ezaki Glico

56

57

57 ツインアーモンド

CD 中田一弘
AD 駒 芳昭
　　金澤一興
D 金澤一興
IL 河又正男
CL 江崎グリコ

Chocolate

CD Kazuhiro Nakada
AD Yoshiaki Koma
　　Kazuoki Kanazawa
D Kazuoki Kanazawa
IL Masao Kawamata
CL Ezaki Glico

58

58 ソリッドグループ

AD 薄井豊子
藤きみよ
D 薄井豊子
藤きみよ
塚原郁乃
IL 脇 庄次
P 黄 建勲
池田 勤
武 典雄
CL 森永製菓

Chocolate

AD Toyoko Usui
Kimiyo Fuji
D Toyoko Usui
Kimiyo Fuji
Ikuno Tsukahara
IL Shoji Waki
P Kenkun Ko
Tsutomu Ikeda
Norio Take
CL Morinaga

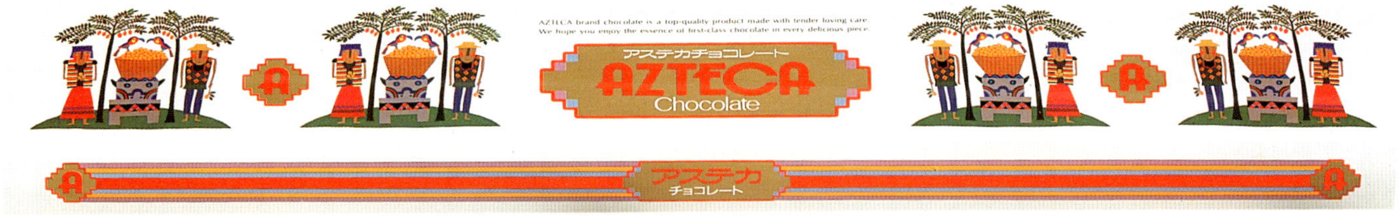

59

59 アスティカ

PL 明治製菓
AD 原田泰治
D 原田泰治
GD 原田泰治
CL 明治製菓
アスティカ

Chocolate

PL Meiji Seika Kaisha
AD Taizi Harada
D Taizi Harada
GD Taizi Harada
CL Meiji Seika Kaisha
Asteca

60

61

62

60 ぬーぼーぼーろ

AD 佐藤勝則
D 福岡加良子
LO 平山博之
CL 森永製菓

Cookies

AD Katsunori Sato
D Kayoko Fukuoka
LO Hiroyuki Hirayama
CL Morinaga

61 トリュフ工房 シュートリュフ

AD 大塚 隆
D 藤きみよ
塚原郁乃
LO 平山博之
P 池田 勉
CL 森永製菓

Chocolate

AD Takashi Otsuka
D Kimiyo Fuji
Ikuno Tsukahara
LO Hiroyuki Hirayama
P Tsutomu Ikeda
CL Morinaga

62 フロスティーナ

PL 桑 和美
AD 桑 和美
D 桑 和美
Const 桑 和美
IL 久保隆夫
P 鈴木裕二
CL 日東コーン

Cake

PL Kazumi Kuwa
AD Kazumi Kuwa
D Kazumi Kuwa
Const Kazumi Kuwa
IL Takao Kubo
P Yuji Suzuki
CL Nitto Cone

63 ジャージーアイスクリーム

AD 出口利春
D 清水幸生
IL 半沢麻男
CL キープ協会

Ice Cream

AD Toshiharu Deguchi
D Yukio Shimizu
IL Asao Hanzawa
CL Kiyosato Educational Experiment Project

63

64

64 フロムザミルク

AD 奥村昭夫
D 大高郁子
CL 江崎グリコ

Ice Cream

AD Akio Okumura
D Ikuko Otaka
CL Ezaki Glico

65

66

66 ヨックモック ショコラ

PL 近藤 司
CD 松原章二
AD 大森幸代
D 大森幸代
高林めぐみ
CL ヨックモック

Chocolate

PL Tsukasa Kondo
CD Shoji Matsubara
AD Yukiko Omori
D Yukiko Omori
Megumi Takabayashi
CL Yoku Moku

65 資生堂パーラー
フルーツゼリー／
クレープショコラ／リーフパイ

AD 仲條正義
D 仲條正義
渋谷善雄
DL 資生堂パーラー

Confectionery

AD Masayoshi Nakajo
D Masayoshi Nakajo
Yoshio Shibuya
DL Shiseido Parlour

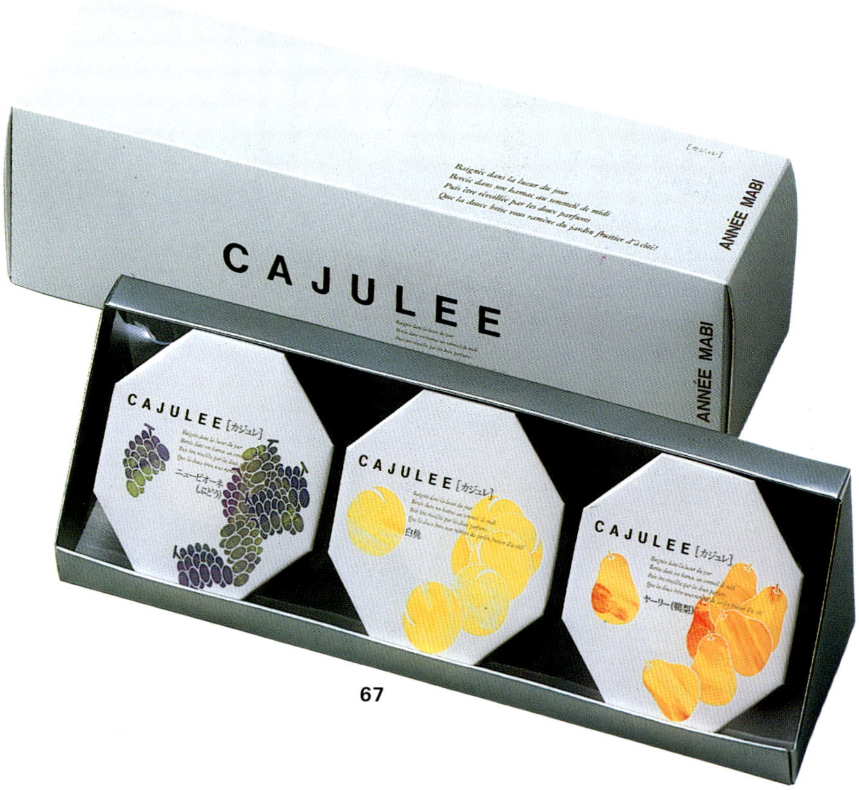

67

67 カジュレ

AD 前田和紀
D 前田和紀
Const 犬塚 保（マックスプランニング）
CL 本高砂屋

Fruit Jelly

AD Kazuki Maeda
D Kazuki Maeda
Const Tamotsu Inuzuka
CL Hontakasagoya

68

68 '92 バレンタイン パリコレクション
AD 池田 毅
D 釣谷 隆
CL ナガサキヤ

Chocolate
AD Takeshi Ikeda
D Takashi Tsuriya
CL Nagasakiya

69 1992年度バレンタイン商品
AD 高橋 篤
D 高橋 篤
　 山下真理子
CL モロゾフ

Chocolate
AD Atsushi Takahashi
D Atsushi Takahashi
　 Mariko Yamashita
CL Morozoff

CHERRY BONBON
BIENAIMÉ
BRANDY BONBON
Lovers can see to do their amorous rites By their own beauties;
or if love be blind, It best agrees with night.— Come, civil night.
BRANDY BONBON
TRUFFES

69

70

70 ラフォーレ
AD 池田 毅
D 釣谷 隆
CL ナガサキヤ

Chocolate
AD Takeshi Ikeda
D Takashi Tsuriya
CL Nagasakiya

71 ローザンヌ
D　渡辺隆史
CL　メリーチョコレートカムパニー

Chocolate
D　Takashi Watanabe
CL　Mary Chocolate

72 ショコラ エ ボンボン
AD　フーシェ・ジャパン
D　矢澤省三
CL　フーシェ・ジャパン

Chocolate/Candy
AD　Foucher Japan
D　Shozo Yazawa
CL　Foucher Japan

71

72

Mary's products have gained a favorable reputation among consumers for their carefully selected ingredients blended into a flavorous taste.
Mary's
LUSTRE
Chocolate
SINCE 1950
Mary's
SPECIALLY SELECTED FINE SWEETS
LUSTRE
Chocolate
SELECTION OF FANCY CHOCOLATES
Mary's products have gained a favorable reputation among consumers for their carefully selected ingredients blended into a flavorous taste. The maker is highly confident that the tasty products of high quality will satisfy consumers.
Mary's

73

74

73 バレンタイン商品 ペイズリーシリーズ

D 渡辺隆史
CL メリーチョコレートカムパニー

Chocolate

D Takashi Watanabe
CL Mary Chocolate

74 ヨックモック クッキーCAN 夏／冬

PL 近藤 司
AD 松原章二
D 川路ヨウセイ
小澤満佐子
大森幸代
IL 松野勇二
殖田綾子
CL ヨックモック

Cookies

PL Tsukasa Kondo
AD Shoji Matsubara
D Yosei Kawaji
Masako Ozawa
Yukiyo Omori
IL Yuji Matsuno
Ayako Ueda
CL Yoku Moku

75

75	フォーチュン チョコクッキー	Cookies	
PL	和田隆次	PL	Ryuji Wada
CD	松浦孝司	CD	Takashi Matsuura
AD	安藤譲治	AD	Joji Ando
D	安藤譲治	D	Joji Ando
	プラウ		The Plow
Const	安藤譲治	Const	Joji Ando
Copy	和田隆次	Copy	Ryuji Wada
CL	高島屋	CL	Takashimaya

76

76 ライラックドリーム&ミスティームーン

AD 町田則一
D 町田則一
IL 手島加江
Copy 町田則一
CL 平塚製菓

Chocolate

AD Norikazu Machida
D Norikazu Machida
IL Kae Tejima
Copy Norikazu Machida
CL Hiratsuka Confectionery

77 ハロウィーン ジャックオランターン／ゴーストン

AD 高橋 篤
D 松井澄江
CL モロゾフ

Candy／Chocolate

AD Atsushi Takahashi
D Sumie Matsui
CL Morozoff

77

78

78 パリパリバー
AD 佐藤勝則
D 佐藤勝則
P 黄 建勲
CL 森永製菓

Ice Cream
AD Katsunori Sato
D Katsunori Sato
P Kenkun Ko
CL Morinaga

79

79 珈琲茶館
エスプレッソコーヒーキャンディ
AD 大久保隆
D 大久保隆
CL カンロ

Candy
AD Takashi Okubo
D Takashi Okubo
CL Kanro

80

80 ブラック&ビター
AD 福岡加良子
D 福岡加良子
P 池田 勉
CL 森永製菓

Ice Cream
AD Kayoko Fukuoka
D Kayoko Fukuoka
P Tsutomu Ikeda
CL Morinaga

81 キスミント

CD 中田一弘
AD 駒 芳昭
金澤 興
D 金澤一興
IL 金澤一興
CL 江崎グリコ

Gam

CD Kazuhiro Nakada
AD Yoshiaki Koma
Kazuaoki Kanazawa
D Kazuoki Kanazawa
IL Kazuoki Kanazawa
CL Ezaki Glico

81

82

82 ミスクリーン

PL 桑 和美
AD 桑 和美
D 桑 和美
CL 日東コーン

Candy

PL Kazumi Kuwa
AD Kazumi Kuwa
D Kazumi Kuwa
CL Nitto Cone

83

83 シャンドーレバレンタイン
'91 '92シリーズ

AD 宇田 優
D 奥田一明
CL シャンドーレ

Chocolate

AD Masaru Uda
D Kazuaki Okuda
CL Chamsdorés

84 ホワイトデイ
AD 森菊五郎
D 下村真弓
IL 梶山明子
CL ベルン

Cake
AD Kikugoro Mori
D Mayumi Shimomura
IL Akiko Kajiyama
CL Berne

84

85

85 世界のお菓子コレクション
AD 森菊五郎
D 下村真弓
IL 梶山明子
CL ベルン

Cookies
AD Kikugoro Mori
D Mayumi Shimomura
IL Akiko Kajiyama
CL Berne

86

86 焼味伝承シリーズ
AD 池田 毅
D 八田隆幸
CL ぼんち

Rice Cracker
AD Takeshi Ikeda
D Takayuki Hachida
CL Bonchi

87 ジャック

AD 鈴木麻子
　 田坂厚生
D 岡田宏三
　 田中春雄
P 橋田憲行
CL ハウス食品工業

Snack

AD Asako Suzuki
　 Atsuo Tasaka
D Kozo Okada
　 Haruo Tanaka
P Noriyuki Hashida
CL Hause Food Industrial

87

88

88 とうもろこしチップス

CD 亀田製菓商品開発部
AD 平本勝彦デザイン室
D 平本勝彦
CL 亀田製菓

Snack

CD Kameda Seika
AD Katsuhiko Hiramoto Design Room
D Katsuhiko Hiramoto
CL Kameda Seika

89 オー・ザック

AD 鈴木麻子
　 奥村昭夫
D 野上周一
P 大石洋一
CL ハウス食品工業

Snack

AD Asako Suzuki
　 Akio Okumura
D Shuichi Nogami
P Yoichi Oishi
CL House Food Industrial

89

90

91

90 テイカロニューフルーツシリーズ
PL アートデザインセンター
D 河田十得
P 高吉幸雄
Copy ジャイ・ナラヤン
CL 三星食品

Candy
PL Art Design Center
D Toru Kawada
P Yukio Takayoshi
Copy Jai Narayan
CL Sansei Foods

91 メープルシロップキャンディ
AD 町田則一
D 町田則一
P 中村新市
Copy 町田則一
CL 平塚製菓

Candy
AD Norikazu Machida
D Norikazu Machida
P Shin'ichi Nakamura
Copy Norikazu Machida
CL Hiratsuka Confectionery

92 純梅キャンデー

AD 小柳登史枝
D アカサカデザインハウス
安上正二
松田茂和
CL ヤマザキ ナビスコ

Candy

AD Toshie Koyanagi
D Akasaka Design House
Masaji Yasugami
Shigekazu Matsuda
CL Yamazaki Nabisco

93 菓匠

AD 千葉和夫
D 土井麻由美
CL ノーベル製菓

Candy

AD Kazuo Chiba
D Mayumi Doi
CL Nobel Confectionery

92

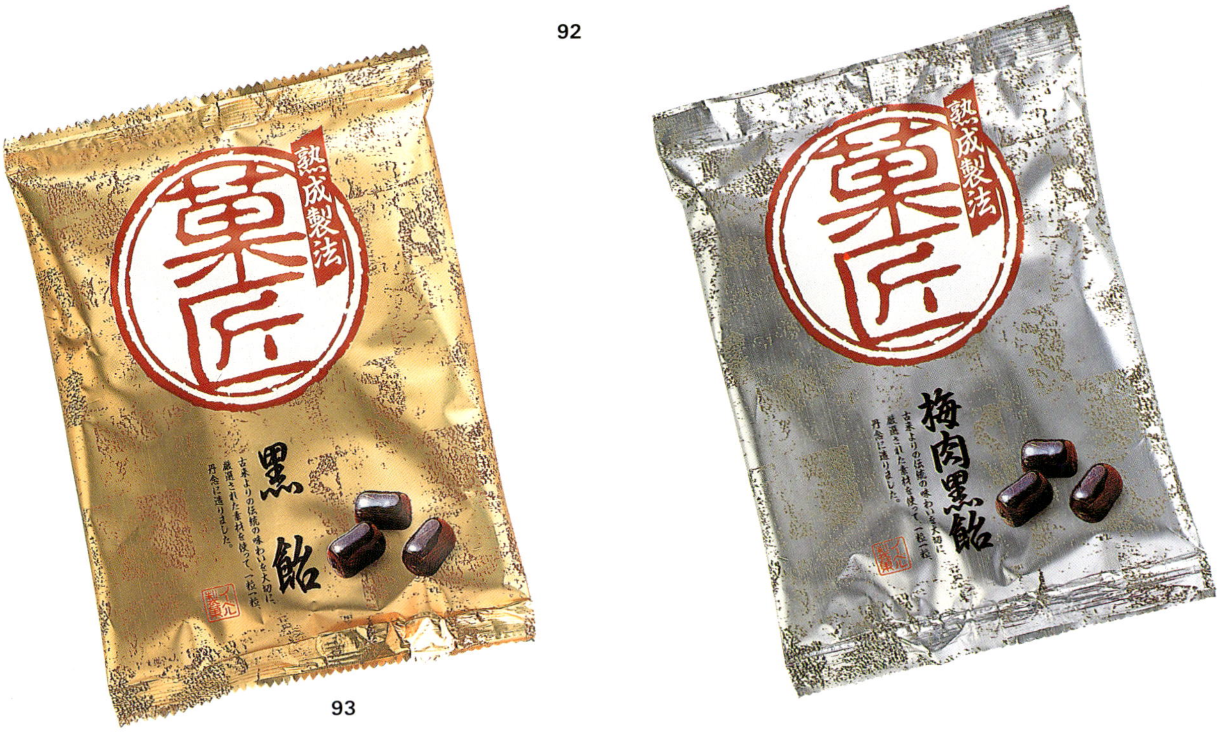

93

94 ママ&エッグス

PL 河田十得
D 河田十得
IL 河田十得
PD 河田十得
CL 三星食品

Candy

PL Toru Kawada
D Toru Kawada
IL Toru Kawada
PD Toru Kawada
CL Sansei Foods

94

95

95	プリン・ア・ラ・モード		**Pudding**
AD	荒木茂野	AD	Shigeno Araki
D	荒木茂野	D	Shigeno Araki
IL	荒木茂野	IL	Shigeno Araki
CL	西洋浪漫菓子 長崎堂	CL	Seiyoromangashi Nagasakido

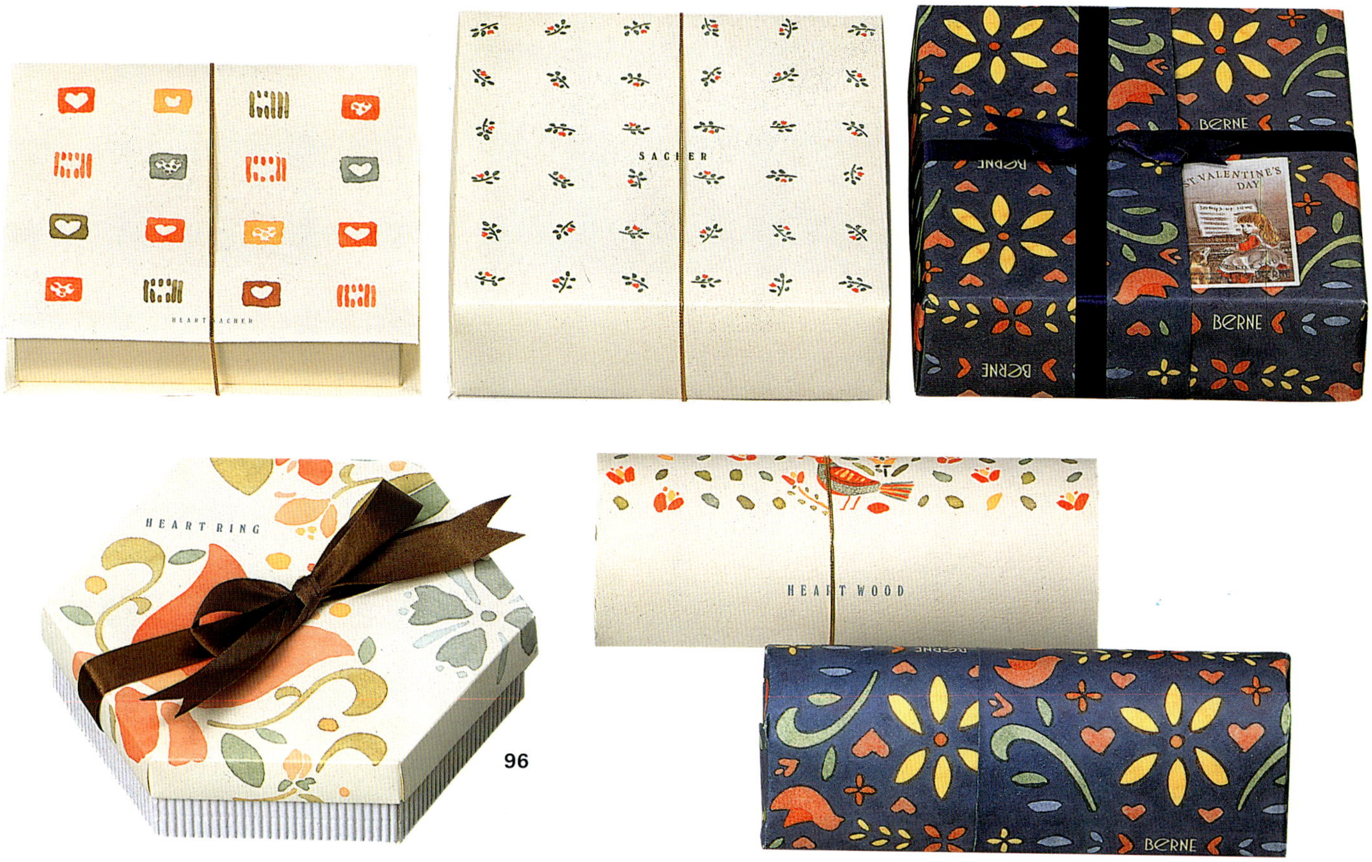

96

96	バレンタインデイ		**Cake**
AD	森菊五郎	AD	Kikugoro Mori
D	森菊五郎	D	Kikugoro Mori
IL	梶山明子	IL	Akiko Kajiyama
CL	ベルン	CL	Berne

97 橘屋 京懐石
AD 神崎 久
D 神崎 久
CL 橘屋

Tea Ceremony Dishes
AD Hisashi Kanzaki
D Hisashi Kanzaki
CL Tachibanaya

97

98

98 笑春風
PL 石田 透
宇佐美彰
AD 石田 透
D 石田 透
CL 樹樹

Japanese Sweets
PL Toru Ishida
Akira Usami
AD Toru Ishida
D Toru Ishida
CL Juju

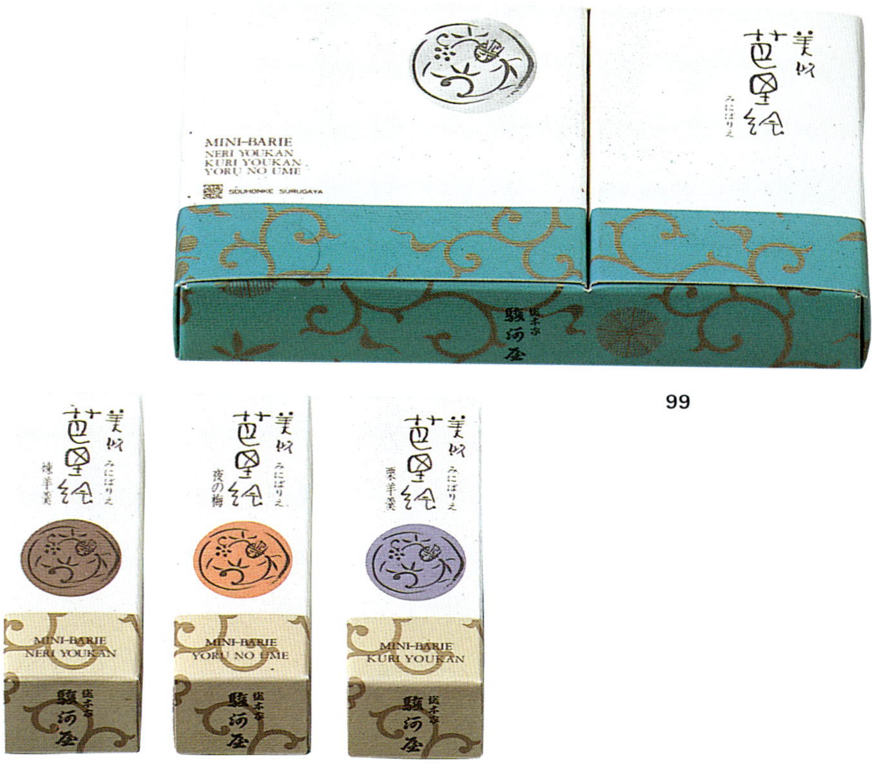

99

99 美似芭里絵

AD 前田和紀
D 前田和紀
山地良子
CA 前田誠一
CL 駿河屋

Japanese Sweets

AD Kazuki Maeda
D Kazuki Maeda
Yoshiko Yamaji
CA Seiichi Maeda
CL Souhonke Surugaya

100 大野屋 栗きんとん

AD 畠山耕治
D 畠山博子
CL 大野屋

Japanese Sweets

AD Koji Hatakeyama
D Hiroko Hatakeyama
CL Ohno ya Confectionery

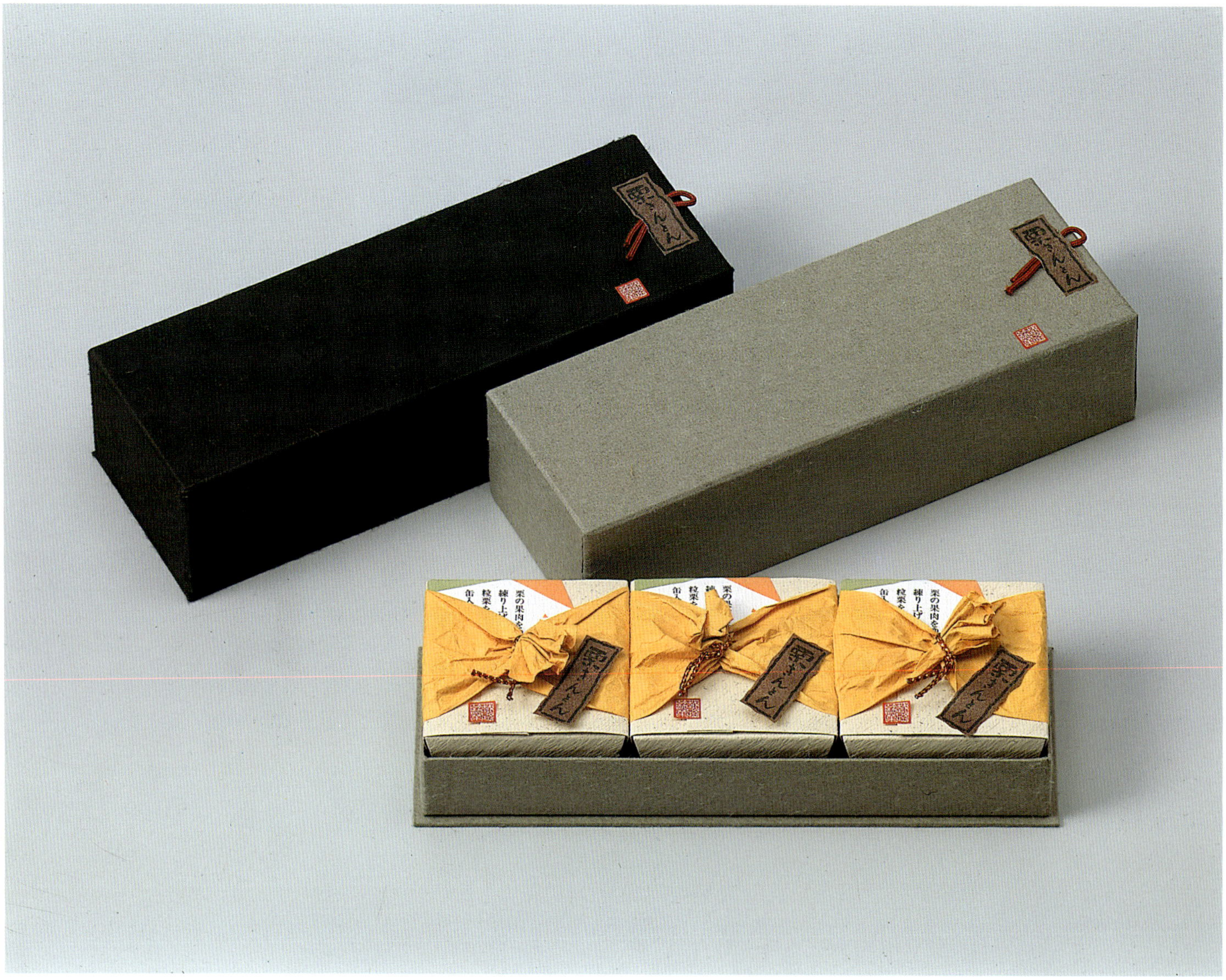

100

101 面野和菓子 紫蘇餅

AD シヴァデザイン事務所
D 清水直明
CL 面野製菓

Japanese Sweets

AD Siva Design Office
D Naoaki Shimizu
CL Omono Confectionery

102 赤米の寿

AD 荒木茂野
D 荒木茂野
IL 荒木茂野
CL 長崎堂

Cake

AD Shigeno Araki
D Shigeno Araki
IL Shigeno Araki
CL Nagasakido

101

102

アルコール飲料　一般飲料

Alcholic Beverage/Soft Drink

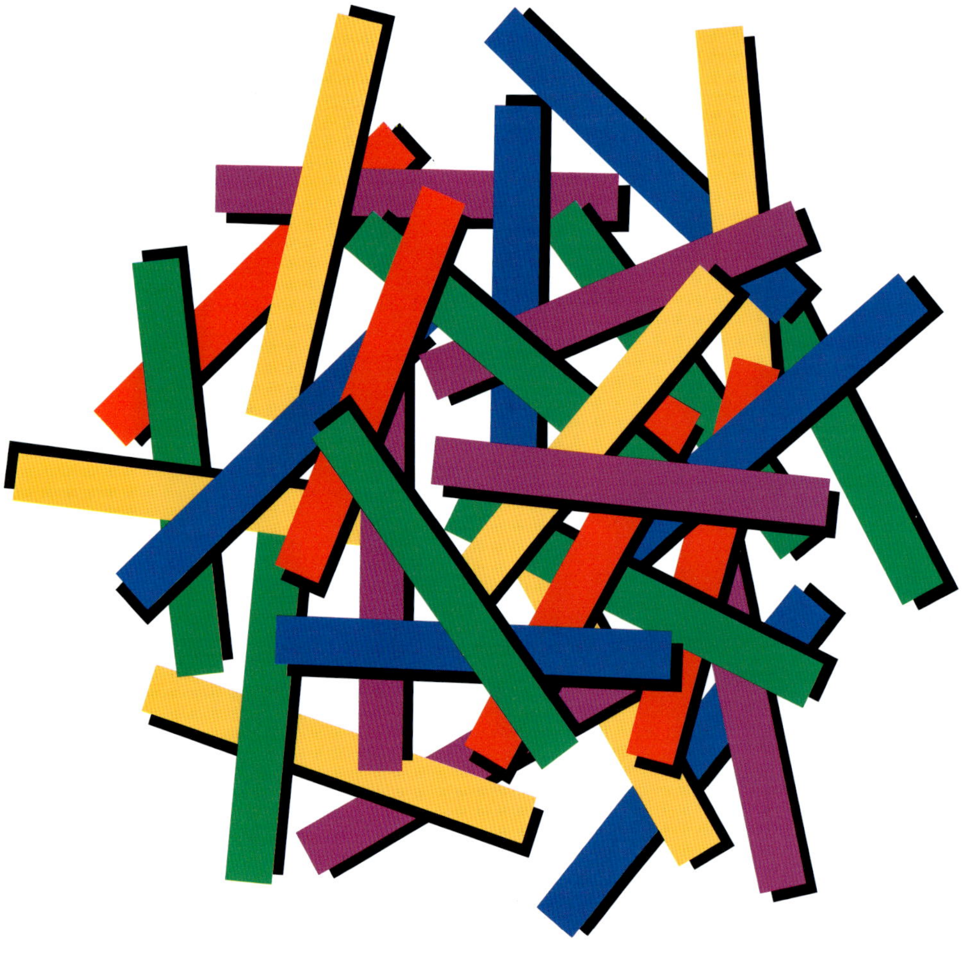

103

104

103 ビア吟生殻やぶり
PL 大森重志
AD 川西 明
D 水口洋二
CL サントリー

Beer
PL Shigeshi Omori
AD Akira Kawanishi
D Yoji Minakuchi
CL Suntory

104 サントリービール モルツ
AD 柳田美保
D 柳田美保
CL サントリー

Beer
AD Miho Yanagida
D Miho Yanagida
CL Suntory

105

106

105 サントリー ビア・ヌーボー1990

AD 高見雄司
D 高見雄司
寿精版印刷デザイン室
CL サントリー

Beer

AD Yuji Takami
D Yuji Takami
Kotobuki Seihan Printing Design Room
CL Suntory

106 アサヒビール ゼット

CD 清野嘉平
AD 佐々木豊
D 佐々木豊
PR 日本デザインセンター
CL アサヒビール

Beer

CD Kahei Kiyono
AD Yutaka Sasaki
D Yutaka Sasaki
PR Nippon Design Center
CL Asahi Breweries

107

108

107 アサヒビール ワイルドビート

CD 清野嘉平
AD 佐々木豊
D 佐々木豊
IL 陰山 肇
PR 日本デザインセンター
CL アサヒビール

Beer

CD Kahei Kiyono
AD Yutaka Sasaki
D Yutaka Sasaki
IL Hajime Kageyama
PR Nippon Design Center
CL Asahi Breweries

108 アサヒビール ほろにが

CD 清野嘉平
AD 佐々木豊
D 佐々木豊
PR 日本デザインセンター
CL アサヒビール

Beer

CD Kahei Kiyono
AD Yutaka Sasaki
D Yutaka Sasaki
PR Nippon Design Center
CL Asahi Breweries

109

109 キリン ゴールデンビター生ビール
CD 眞木 準
AD 戸田正寿
D 鈴木 守
CL キリンビール

Beer
CD Jun Maki
AD Masatoshi Toda
D Mamoru Suzuki
CL Kirin Brewery

KIRIN BREWERY COMPANY, LIMITED
KIRIN
LAGER
BEER
キリンラガービール
500ml
1000ml
350ml

110

111 サントリービール 夏の生 1992
AD 大門敏彦
D ケン・ドーン
CL サントリー

Beer
AD Toshihiko Daimon
D Ken Done
CL Suntory

112 カルピスフィズ
AD 桑 和美
D 桑 和美
CL カルピス食品工業

Lactic Acid Cocktail
AD Kazumi Kuwa
D Kazumi Kuwa
CL The Calpis Food Industry

111

112

110 キリンラガービール
AD 松永 真
D 松永 真
CL キリンビール

Beer
AD Shin Matsunaga
D Shin Matsunaga
CL Kirin Brewery

113

113 サントリー フルーツ&ビア
AD 牛島志津子
D 牛島志津子
大平尚子
CL サントリー

Fruit/Beer
AD Shizuko Ushijima
D Shizuko Ushijima
Naoko Ohira
CL Suntory

114

115

114 サントリー 紅茶のお酒

AD 加藤芳夫
D 柴戸由岐子
加藤芳夫
CL サントリー

Liqueur

AD Yoshio Kato
D Yukiko Shibato
Yoshio Kato
CL Suntory

115 サントリー お菓子なお酒

AD 加藤芳夫
D 柴戸由岐子
IL 大沼 透
CL サントリー

Liqueur

AD Yoshio Kato
D Yukiko Shibato
IL Toru Onuma
CL Suntory

116 サントリーピュアモルトウイスキー
山崎(アコーディオン型ボトル)
サントリーブランデーエクストラ
(マリンバ型ボトル)
サントリーウイスキー
エイジング**15**(ハープ型ボトル)

AD 横尾 浩
牛島志津子
D 牛島志津子
CL サントリー

Whisky/Brandy

AD Hiroshi Yokoo
Shizuko Ushijima
D Shizuko Ushijima
CL Suntory

SUNTORY
AGEING
15 YEARS
ウイスキー
SUNTORY WHISKY
AGEING-15-
ウイスキー

116

117

118

117 ニッカ ピュアモルト・カフェグレーン

CD 清野嘉平
AD 原 研哉
D 原 研哉
PR 日本デザインセンター
CL ニッカウヰスキー

Whisky

CD Kahei Kiyono
AD Kenya Hara
D Kenya Hara
PR Nippon Design Center
CL The Nikka Whisky Distilling

118 サントリーブランデーV. S. O. P
スリムボトル

AD 大門敏彦
D 森 旬子
CL サントリー

Brandy

AD Toshihiko Daimon
D Junko Mori
CL Suntory

119 キリン・シーグラム ウイスキー
ヒップス

PL 浅野高弘
AD 小南 潔
D 小南 潔
高橋丈治
IL 松里真守
Copy 小池光洋
CL キリン・シーグラム

Whisky

PL Takahiro Asano
AD Kiyoshi Kominami
D Kiyoshi Kominami
Joji Takahashi
IL Mamoru Matsuzato
Copy Mitsuhiro Koike
CL Kirin Seagram

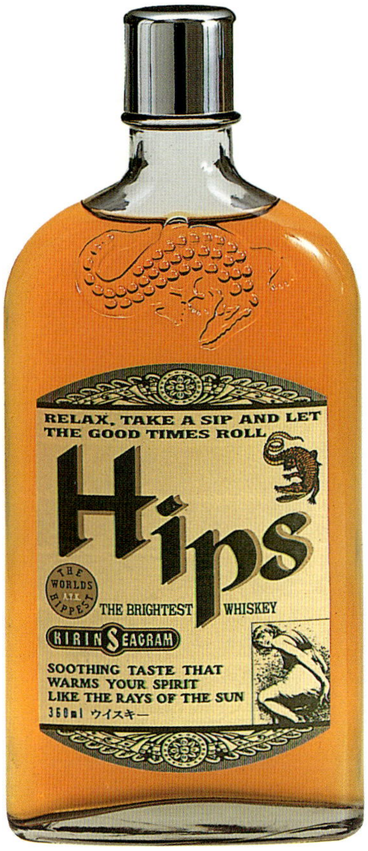

119

120

120 シングルトン

AD 久保田利通
D 檜山永次
Const 檜山永次
CL 日本ヒューブライン

Whisky

AD Toshimichi Kubota
D Eiji Hiyama
Const Eiji Hiyama
CL Heublein

121

121	**華月**	**Sake**	
AD	奥村昭夫	AD	Akio Okumura
D	竹村真澄	D	Masumi Takemura
LO	荻野丹雪	LO	Tansetsu Ogino
CL	月桂冠	CL	Gekkeikan

122

122 サントリー ピュアモルトウイスキー
木桶仕込 古樽仕上

AD 大門敏彦
浅倉 正
D 大門敏彦
浅倉 正
LO 荻野丹雪
IL 吉田勝彦
CL サントリー

Whisky

AD Toshihiko Daimon
Tadashi Asakura
D Toshihiko Daimon
Tadashi Asakura
LO Tansetsu Ogino
IL Katsuhiko Yoshida
CL Suntory

純
宝焼酎
富士山麓湧水使用
焼酎甲類25度720ml
15th
ANNIVERSARY
純
15th
ANNIVERSARY

123

124

123 宝焼酎 純 15周年記念ボトル

PL ユニゾンネットワーク
AD 佐藤 卓
D 佐藤 卓
CL 宝酒造

Spirit

PL Unison Network
AD Taku Sato
D Taku Sato
CL Takara Shuzo

124 十八代伊兵衛

CD 阿部憲人
AD 竹智 淳
D 竹智 淳
CL 奥の松酒造

Sake

CD Norihito Abe
AD Jun Takechi
D Jun Takechi
CL Okunomatsu Shuzo

媛の愛
天味・純米大吟醸
美味・純米吟醸
雪雀
清酒
YUKISUZUME
幻味・純米大吟醸
梅錦
清酒
UMENISHIKI
天味・純米大吟醸
江戸錦
清酒
EDONISHIKI

125

125 媛の愛
AD 松井桂三
D 松井桂三
荒木優子
CL 愛媛県酒造組合

Sake
AD Keizo Matsui
D Keizo Matsui
Yuko Araki
CL The Brewery Coop of Ehime Prefecture

126 吟醸生貯蔵酒
AD TCD
D 神谷利男
星加 健
Copy 手束 卓
CL 白鶴酒造

Sake
AD TCD
D Toshio Kamitani
Takeshi Hoshika
Copy Takashi Tezuka
CL Hakutsuru Sake Brewing

127 日本酒 わしが国
AD 戸田正寿
D 鈴木 守
CL 山和酒造店

Sake
AD Masatoshi Toda
D Mamoru Suzuki
CL Sanwa Sake Brewing

126

127

128 本醸造清酒 熊楠

PL 武友正修
AD 武友正修
D 武友朋子
Copy 武友正修
IL 南方熊楠
CL 世界一統

Sake

PL Masanobu Taketomo
AD Masanobu Taketomo
D Tomoko Taketomo
Copy Masanobu Taketomo
IL Kumagusu Minakata
CL Sekaiitto

128

129

129 麦焼酎 白水

AD 近藤洋太郎
村 禎介
D 伊東 保
IL 伊東 保
CL メルシャン

Shochu

AD Yotaro Kondo
Teisuke Mura
D Tamotsu Ito
IL Tamotsu Ito
CL Mercian

130

131

130 白鶴のみくらべ

AD TCD
D 神谷利男
田端 恵
Copy 手束 卓
IL 織田イラストレーション事務所
CL 白鶴酒造

Sake

AD TCD
D Toshio Kamitani
Megu Tabata
Copy Takashi Tezuka
IL Oda Illustration Office
CL Hakutsuru Sake Brewing

131 ユーコープ焼酎シリーズ

CD 横田 誠
AD 坪谷一利
D 坪谷一利
CL かながわ生協

Shochu

CD Makoto Yokota
AD Kazutoshi Tsuboya
D Kazutoshi Tsuboya
CL Kanagawa Coop

132

132 スズキコージ イラストボトル

PL 翔泳社デザイン研究所
AD 篠﨑晃一
D 佐藤直樹
IL スズキコージ
CL 勝沼醸造

Wine

PL Shoeisha Design Lab.
AD Koichi Shinozaki
D Naoki Sato
IL Koji Suzuki
CL Katsunuma Winery

133

133 小さな葡萄酒

AD 藤田 隆
D 森 旬子
CL サントリー

Wine

AD Takashi Fujita
D Junko Mori
CL Suntory

134

134 サントリーりんごのお酒
ぶどうのお酒

AD 藤田 隆
牛島志津子
D 牛島志津子
CL サントリー

Wine

AD Takashi Fujita
Shizuko Ushijima
D Shizuko Ushijima
CL Suntory

135

135 サントネージュワイン 甲斐の国

AD 橋本近雄
D 新保欣也
真藤 哲
CL 協和発酵工業

Wine

AD Chikao Hashimoto
D Kinya Shimbo
Akira Shindo
CL Kyowa Hakko Kogyo

136

136 オーストラリアワイン サザン・クロス

PL 大森重志
AD 前田英樹
D 前田英樹
寿精版印刷デザイン室
CL サントリー

Wine

PL Shigeshi Omori
AD Hideki Maeda
D Hideki Maeda
Kotobuki Seihan Printing Design Room
CL Suntory

137 サントリーワイン マスカット

AD 吉田勝彦
D 吉田勝彦
IL 吉田勝彦
CL サントリー

Wine

AD Katsuhiko Yoshida
D Katsuhiko Yoshida
IL Katsuhiko Yoshida
CL Suntory

138 キリン ワインクラブ ダンス

AD 佐藤昭夫
D 佐藤昭夫
IL 佐藤昭夫
CL キリンビール

Wine

AD Akio Sato
D Akio Sato
IL Akio Sato
CL Kirin Brewery

137

138

139 二十四節季酒

PL　フォアトップス
AD　塩沢喜徳
D　塩沢喜徳
AG　フォアトップス
CL　御代桜醸造
小澤酒造
辻本店
西武百貨店

Sake

PL　Four Tops
AD　Yoshinori Shiozawa
D　Yoshinori Shiozawa
AG　Four Tops
CL　Miyozakura Jozo
Ozawa Shuzo
Tsuji Honten
Seibu Department Store

139

140 ピーチワイン／ストロベリーワイン／グランドメロンワイン

AD　榊原勝一
D　榊原勝一
IL　榊原勝一
CL　はこだてわいん

Wine

AD　Katsuichi Sakakibara
D　Katsuichi Sakakibara
IL　Katsuichi Sakakibara
CL　Hakodate Winery

140

141

141 マキシムエステートレギュラーコーヒー

CD 田島 肇
AD 河北秀也
D 谷井郁美
CL 味の素

Coffee

CD Hajime Tajima
AD Hideya Kawakita
D Yubi Tanii
CL Ajinomoto

142 カフェ・デ・コロンビア クリストバル

PL 渡邊潤一郎
AD GKグラフィックス
D GKグラフィックス
CL 日本紅茶

Coffee

PL Jun'ichiro Watanabe
AD GK Graphics
D GK Graphics
CL Japan Black Tea

Café de Colombia
CRISTÓBAL
Supremo #18
CRISTÓBAL
tostado al carbón
Café de Colombia
CRISTÓBAL
NARIÑO dosmil
#16
Excelso
Café de Colombia
CRISTÓBAL
SANTANDER
#18
Supremo
Café de Colombia
CRISTÓBAL
CRISTÓBAL
#16
Excelso

142

143

143 プレッソ プレッソ
PL ネッスル日本
AD 須田清次
D 阿部恭子
前原えみか
CL ネッスル日本

Coffee
PL Nestle Japan
AD Seiji Suda
D Kyoko Abe
Emika Maehara
CL Nestle

144 新グランデージ
スーパープレミアムコーヒー
CD 田島 肇
AD 本田和男
D 本田和男
水上千文
CL 味の素ゼネラルフーヅ

Coffee
CD Hajime Tajima
AD Kazuo Honda
D Kazuo Honda
Chifumi Mizukami
CL Ajinomoto General Foods

144

145 マスターズコレクション ミックススリー
AD 河北秀也
D 明星秀隆
谷井郁美
CL 味の素ゼネラルフーヅ

Coffee

AD Hideya Kawakita
D Hidetaka Myojo
Youbi Tanii
CL Ajinomoto General Foods

145

146 新マキシム
AD 原 研哉
D 原 研哉
井上幸恵
AC 田島 肇
CL 味の素ゼネラルフーヅ

Coffee

AD Kenya Hara
D Kenya Hara
Yukie Inoue
AC Hajime Tajima
CL Ajinomoto General Foods

146

147

147 マキシムリキッドコーヒー

AD　本田和男
D　本田和男
　　水上千文
CL　味の素ゼネラルフーヅ

Coffee

AD　Kazuo Honda
D　Kazuo Honda
　　Chifumi Mizukami
CL　Ajinomoto General Foods

148 ブレンディ缶コーヒー

AD 松永 真
D 松永 真
鳩貝浩司
CL 味の素ゼネラルフーヅ

Coffee

AD Shin Matsunaga
D Shin Matsunaga
Koji Hatogai
CL Ajinomoto General Foods

149 ブレンディボトルコーヒー

AD 松永 真
D 松永 真
鳩貝浩司
CL 味の素ゼネラルフーヅ

Coffee

AD Shin Matsunaga
D Shin Matsunaga
Koji Hatogai
CL Ajinomoto General Foods

148

149

150

150 アサヒコーヒー J.O.

CD 清野嘉平
AD 久保 亨
D 久保 亨
PR 日本デザインセンター
CL アサヒビール飲料

Coffee

CD Kahei Kiyono
AD Toru Kubo
D Toru Kubo
PR Nippon Design Center
CL Asahi Beverage International

151 サントリーコーヒー ボス

CD 大門敏彦
AD 加藤芳夫
D 石浦弘幸
CL サントリー

Coffee

CD Toshihiko Daimon
AD Yoshio Kato
D Hiroyuki Ishiura
CL Suntory

151

152

153

152 カリビアンコーヒー
CD　和田 亨
D　日下部治朗
IL　林 幸蔵
CL　グリコ協同乳業

Coffee
CD　Toru Wada
D　Jiro Kusakabe
IL　Kozo Hayashi
CL　Glico Dairy

153 モカ&ブルーマウンテン コーヒーEX
CD　和田 亨
D　日下部治朗
CL　グリコ協同乳業

Coffee
CD　Toru Wada
D　Jiro Kusakabe
CL　Glico Dairy

154 マリーム クレミュー
AD　松永 真
D　松永 真
　　鳩貝浩司
CL　味の素ゼネラルフーヅ

Creaming Powder
AD　Shin Matsunaga
D　Shin Matsunaga
　　Koji Hatogai
CL　Ajinomoto General Foods

154

155

156

155 ティークオリティ

CD 清野嘉平
AD 和田マコト
D 和田マコト
PR 日本デザインセンター
CL アサヒビール飲料

Tea

CD Kahei Seino
AD Makoto Wada
D Makoto Wada
PR Nippon Design Center
CL Asahi Beverage International

156 紅茶の樹 もも紅茶／ゆず紅茶／りんご紅茶

CD 大門敏彦
AD 加藤芳夫
D 水口洋二
CL サントリー

Tea

CD Toshihiko Daimon
AD Yoshio Kato
D Yoji Minakuchi
CL Suntory

157 ファーマーズテーブルの紅茶
PL　石川博子
AD　石川 源
D　秋山香代子
CL　ファーマーズテーブル

Tea
PL　Hiroko Ishikawa
AD　Minato Ishikawa
D　Kayoko Akiyama
CL　Farmer's Table

157

158

158 日東ダージリン レギュラーティー
CD　渡辺正久
AD　佐藤 卓
D　佐藤 卓
CL　三井農林

Tea
CD　Masahisa Watanabe
AD　Taku Sato
D　Taku Sato
CL　Mitsui Norin

159 日東セイロン（ディンブラ）
レギュラーティー
CD　渡辺正久
AD　佐藤 卓
D　佐藤 卓
CL　三井農林

Tea
CD　Masahisa Watanabe
AD　Taku Sato
D　Taku Sato
CL　Mitsui Norin

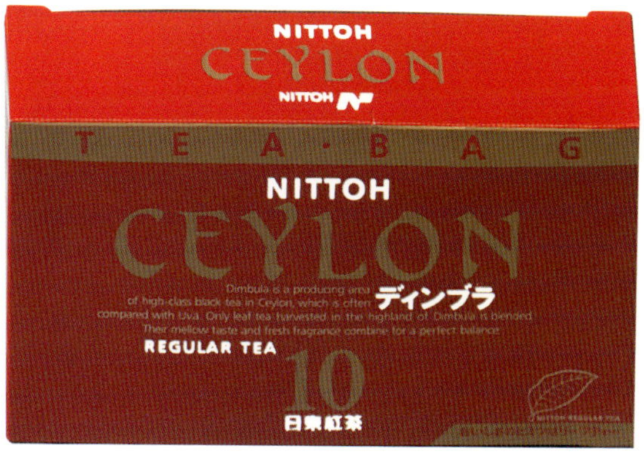

159

160

160 うめスカッシュ
CD 越知 準
AD 犬塚達美
D 犬塚達美
CL 不二家

Soft Drink
CD Jun Ochi
AD Tatsumi Inuzuka
D Tatsumi Inuzuka
CL Fujiya

161 カルピスソーダ
AD 佐藤 卓
D 佐藤 卓
CL カルピス食品工業

Soft Drink
AD Taku Sato
D Taku Sato
CL The Calpis Food Industry

161

162 カルピス オ・レ ライト

CD　和田 亨
D　阿部 優
　　中田淑子
CL　カルピスフーズサービス

Soft Drink

CD　Toru Wada
D　Masaru Abe
　　Yoshiko Nakada
CL　Calpis Food Service

163 カルピス&ミルク

CD　和田 亨
D　中田淑子
　　阿部 優
CL　カルピスフーズサービス

Soft Drink

CD　Toru Wada
D　Yoshiko Nakada
　　Masaru Abe
CL　Calpis Food Service

162

163

164

165

164 キリンレモン セレクト

AD 矢沢孝嗣
D 伊東美保
CL キリンビバレッジ

Soft Drink

AD Takatsugu Yazawa
D Miho Ito
CL Kirin Beverage

165 キリンエスパシオ

PL 亀ヶ谷みほ
CD 橋本近雄
AD 菊地 学
D 菊地 学
CL キリンビバレッジ

Soft Drink

PL Miho Kamegaya
CD Chikao Hashimoto
AD Manabu Kikuchi
D Manabu Kikuchi
CL Kirin Beverage

166 ポッカストレートオレンジジュース/ツリートップストレートアップルジュース/オーシャンスプレープレミアムストレートグレープフルーツジュース

D 久松章男
CL ポッカコーポレーション

Soft Drink

D Akio Hisamatsu
CL Pokka Corporation

167 明治オレンジ100/アップル100

PL 時政美枝
CD 時政美枝
AD 金澤一興
D 金澤一興
IL 金澤一興
P 内山広文
CL 明治乳業

Juice

PL Miki Tokimasa
CD Miki Tokimasa
AD Kazuoki Kanazawa
D Kazuoki Kanazawa
IL Kazuoki Kanazawa
P Hirofumi Uchiyama
CL Meiji Milk Products

TREE TOP
APPLE
アップル 果汁100%
Straight Juice
Squeezed from the highest quality fruit, TREE TOP presents this extravagant 100% apple juice.
POKKA
ORANGE
オレンジ 果汁100%
Straight Juice
Squeezed from the highest quality fruit, POKKA presents this extravagant 100% orange juice.
Ocean spray
GRAPEFRUIT
グレープフルーツ 果汁100%
Straight Juice
Squeezed from the highest quality fruit, OCEAN SPRAY presents this extravagant 100% grapefruit juice.

166

はじめての果汁100%
ベースの乳酸菌飲料
明治アップル
100
MEIJI
93. 3.13 製造
果汁100%
種類別 乳酸菌飲料

167

168 オーシャンスプレープレミアム
グレープフルーツジュース
200g·800g

D　久松章男
CL　ポッカコーポレーション

Soft Drink

D　Akio Hisamatsu
CL　Pokka Corporation

169 カルピス B&L

CD　和田 亨
D　阿部 優
　中田淑子
CL　カルピス食品工業

Soft Drink

CD　Toru Wada
D　Masaru Abe
　Yoshiko Nakada
CL　The Calpis Food Industry

168

169

170 アサヒ完熟ストレート

CD　清野嘉平
AD　久保 亨
D　久保 亨
PR　日本デザインセンター
CL　アサヒビール飲料

Juice

CD　Kahei Kiyono
AD　Toru Kubo
D　Toru Kubo
PR　Nippon Design Center
CL　Asahi Beverage International

170

Good Morning
Café de Café
news paper café
Deliver several types of coffee every month. Why don't you enjoy your luxury coffee break?
Café de Café
Deliver several types of coffee every month.
NEWS PAPER CAFÉ
Another Moon Café
Here is a rich cup of senshukai's original coffee when a beautiful time slowly starts to flow.
BRAZIL BLEND
Another Moon Café
Here is a rich cup of senshukai's original coffee when a beautiful time slowly starts to flow.
NET WEIGHT (90 GRAMS)
LakeSide café
LakeSide café
Café
Café de Café
lake side café
Café de Café
café on the sofa
Deliver several types of coffee every month. Why don't you enjoy your luxury coffee break?
[MOCHA]
Café On The Sofa
Here is a rich cup of senshukai's original coffee—when a beautiful time slowly starts to flow.
[MOCHA]
Café On The Sofa
Here is a rich cup of senshukai's original coffee—when a beautiful time slowly starts to flow.
High Roast

171

171 カフェ・ド・カフェ

AD　田口 圭
D　田口 圭
IL　田口 圭
CL　千趣会

Coffee

AD　Kei Taguchi
D　Kei Taguchi
IL　Kei Taguchi
CL　Senshukai

172 アラブコーヒーパッケージ

AD　田中康夫
D　田中康夫
GD　田中康夫
CL　アラブ珈琲
日本インターナショナル
エージェンシー

Coffee

AD　Yasuo Tanaka
D　Yasuo Tanaka
GD　Yasuo Tanaka
CL　Arab Coffee
NIA

172

174 サントリー 熱血飲料

CD 大門敏彦
AD 加藤芳夫
D 水口洋二
CL サントリー

Nutritive Drink

CD Toshihiko Daimon
AD Yoshio Kato
D Yoji Minakuchi
CL Suntory

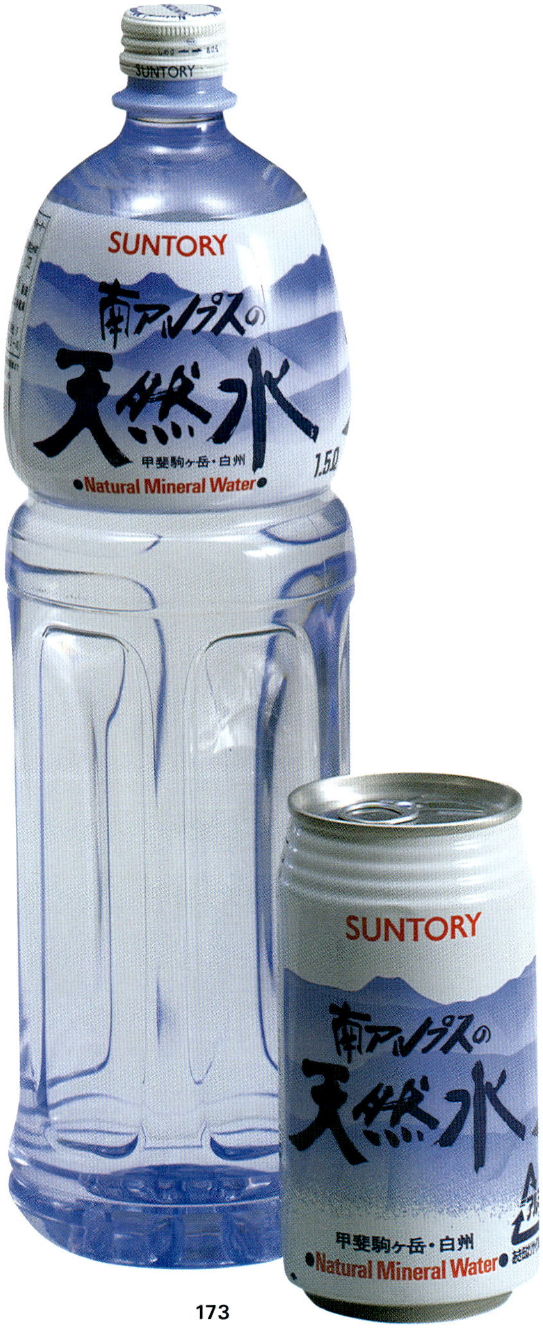

173

174

173 サントリー ナチュラルミネラルウォーター
南アルプスの天然水

CD 大門敏彦
AD 加藤芳夫
池田 毅
D 大久保隆
釣谷 隆
LO 荻野丹雪
CL サントリー

Mineral Water

CD Toshihiko Daimon
AD Yoshio Kato
Takeshi Ikeda
D Takashi Okubo
Takashi Tsuriya
LO Tansetsu Ogino
CL Suntory

175 キリン ポストウォーター

AD 佐藤昭夫
D 佐藤昭夫
BD 日本硝子
東洋ガラス
IL 佐藤昭夫
CL キリンビバレッジ

Hypotonic Drink

AD Akio Sato
D Akio Sato
BD Nippon Glass
Toyo Glass
IL Akio Sato
CL Kirin Beverage Corporation

175

化粧品・香水 化粧雑貨

Cosmetics, Perfume/Toiletry

176

176 コスメデコルテ ポイントメイクシリーズ

CD 中山紀子
AD 塙冨士雄
D 井口佳己
渡辺 均
鈴木智晴
CL コーセー

Cosmetics

CD Michiko Nakayama
AD Fujio Hanawa
D Yoshimi Iguchi
Hitoshi Watanabe
Chiharu Suzuki
CL Kosé

177

177 エレナージュ

CD 大浜良友
AD 梅沢一徳
D 梅沢一徳
CL エイボン・プロダクツ

Cosmetics

CD Yoshitomo Ohama
AD Kazunori Umezawa
D Kazunori Umezawa
CL Avon Products

178

178 アルビオン エクセレント
CV ホワイトⅡ

CD　錦織章三
　　野田徳雄
D　中野 恵
　　佐山和恵
CL　アルビオン

Cosmetics

AD　Shozo Nishigori
　　Norio Noda
D　Megumi Nakano
　　Kazue Sayama
CL　Albion

179

179 ブルーローズ

AD　池田修一
D　工藤青石
IL　平野敬子
CL　資生堂

Fragrance

AD　Shuichi Ikeda
D　Aoshi Kudo
IL　Keiko Hirano
CL　Shiseido

180 メイベリン メイクアップシリーズ
AD　佐藤 卓
D　佐藤 卓
CL　メイベリン

Cosmetics
AD　Taku Sato
D　Taku Sato
CL　Maybelline

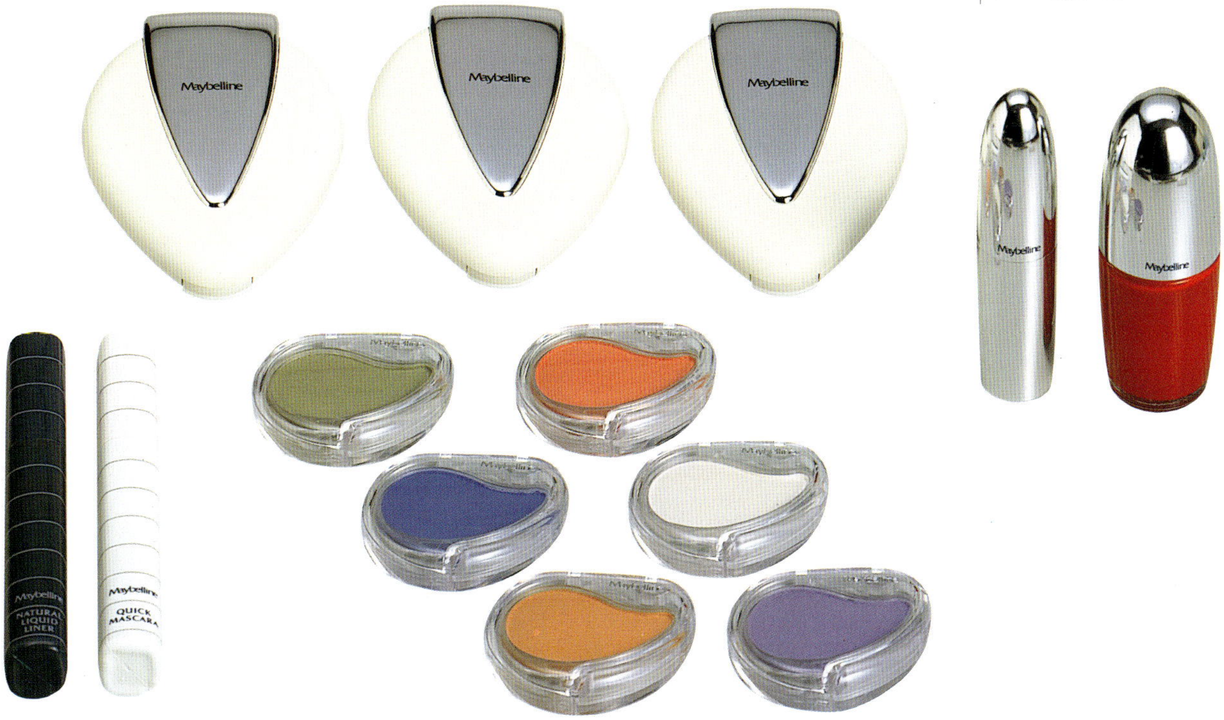

180

181

181	ディシラ	**Cosmetics**	
CAD	花内 勇	CAD	Isamu Hanauchi
AD	池田修一	AD	Shuichi Ikeda
D	廣 哲夫 斗ヶ沢哲雄	D	Tetsuo Hiro Tetsuo Togasawa
LO	ティム・ガービン	LO	Tim Girvin
CL	資生堂	CL	Shiseido

182

182 オブラン
CD　植村 秀
AD　中塚大輔
D　石橋政美
CL　シュウウエムラ化粧品

Cosmetics
CD　Shu Uemura
AD　Daisuke Nakatsuka
D　Masami Ishibashi
CL　Shu Uemura Cosmetics

183

183 バイタルパーフェクション
AD 杉浦俊作
D 塩川 稔
笹原浩造
CL 資生堂

Cosmetics
AD Shunsaku Sugiura
D Minoru Shiokawa
Kozo Sasahara
CL Shiseido

184

184 インウイ
AD 池田修一
D ヴァレリー・ベルナール
陣内昭子
LD カレ・ノアール
CL 資生堂

Cosmetics
AD Shuichi Ikeda
D Valerie Bernard
Akiko Jinnai
LD Carré Noir
CL Shiseido

185

185 アンジェリーク

AD 池田修一
D 工藤青石
LD 博報堂
IL 工藤青石
アンディ・ウォーホル
CL 資生堂

Fragrance

AD Shuichi Ikeda
D Aoshi Kudo
LD Hakuhodo
IL Aoshi Kudo
Andy Warhol
CL Shiseido

186

186 ソニアリキエル
ナイトアンドデイコンパクト
クールドール

AD 錦織章三
D 野田徳雄
篠崎雅子
CL アルビオン

Cosmetics

AD Shozo Nishigori
D Norio Noda
Masako Shinozaki
CL Albion

188

187 モイストウォーター
CAD 太田雅雄
AD 中井一郎
D 磯貝典子
CL 資生堂

Cosmetics
CAD Masao Ota
AD Ichiro Nakai
D Noriko Isogai
CL Shiseido

188 ドゥ·セーズポイントメイク
CD 中山紀子
AD 塙冨士雄
D 井口佳己
渡辺 均
鈴木智晴
CL コーセー

Cosmetics
CD Michiko Nakayama
AD Fujio Hanawa
D Yoshimi Iguchi
Hitoshi Watanabe
Chiharu Suzuki
CL Kosé

187

189 **MOAリマベール**

AD	遠藤紀雄
D	遠藤紀雄
LO	松本恭典
IL	遠藤紀雄
CL	リマナチュラル

Cosmetics

AD	Norio Endo
D	Norio Endo
LO	Kyosuke Matsumoto
IL	Norio Endo
CL	Lima Nutural

189

190

190 **アンテリージェ トリートメントシリーズ**

CD	中山紀子
AD	塙冨士雄
D	林 千博
	乗富茂美
	新鞍一裕
	米徳一葉
	水野悦子
CL	コーセー

Beauty Products

CD	Michiko Nakayama
AD	Fujio Hanawa
D	Chihiro Hayashi
	Shigemi Noritomi
	Kazuhiro Niikura
	Kazuha Yonetoku
	Etsuko Mizuno
CL	Kosé

191 アンテリージェ サマーベース
メイクアップシリーズ

CD 中山紀子
AD 塙冨士雄
D 林 千博
石橋利江子
米徳一葉
CL コーセー

Cosmetics

CD Michiko Nakayama
AD Fujio Hanawa
D Chihiro Hayashi
Rieko Ishibashi
Kazuha Yonetoku
CL Kosé

191

192 アルビオン アプセア

AD 錦織章三
D 佐山和恵
CL アルビオン

Cosmetics

AD Shozo Nishigori
D Kazue Sayama
CL Albion

192

193 エテュセ

AD 斗ヶ沢哲雄
D 笹原浩造
四田千絵子
LO 平戸絵里子
CL エテュセ

Cosmetics

AD Tetsuo Togasawa
D Kozo Sasahara
Chieko Yotsuda
LO Eriko Hirato
CL ET TU SAIS

193

194 アネッサ

AD 池田修一
D 松本 泉
IL 松本 泉
CL 資生堂

Sunscreen Products

AD Shuichi Ikeda
D Izumi Matsumoto
IL Izumi Matsumoto
CL Shiseido

194

195

195 ディニュー スキンケアシリーズ

AD 大野淳一郎
D 碓井健司
宮武かおる
有田 賢
CL ポーラ化粧品本舗

Cosmetics

AD Jun'ichiro Ono
D Takeshi Usui
Kaoru Miyatake
Satoshi Arita
CL Pola

196

196 エリクシール

AD 廣 哲夫
D 松本 泉
鐘尾 隆
CL 資生堂

Cosmetics

AD Tetsuo Hiro
D Izumi Matsumoto
Takashi Kaneo
CL Shiseido

197

197 ルシェリ ベースメイクシリーズ

CD 中山紀子
AD 塙冨士雄
D 井口佳己
矢島 修
鈴木智晴
CL コーセー

Cosmetics

CD Michiko Nakayama
AD Fujio Hanawa
D Yoshimi Iguchi
Osamu Yajima
Chiharu Suzuki
CL Kosé

198

198 エリクシール
AD 廣 哲夫
D 松本 泉
加藤喜代子
CL 資生堂

Cosmetics
AD Tetsuo Hiro
D Izumi Matsumoto
Kiyoko Kato
CL Shiseido

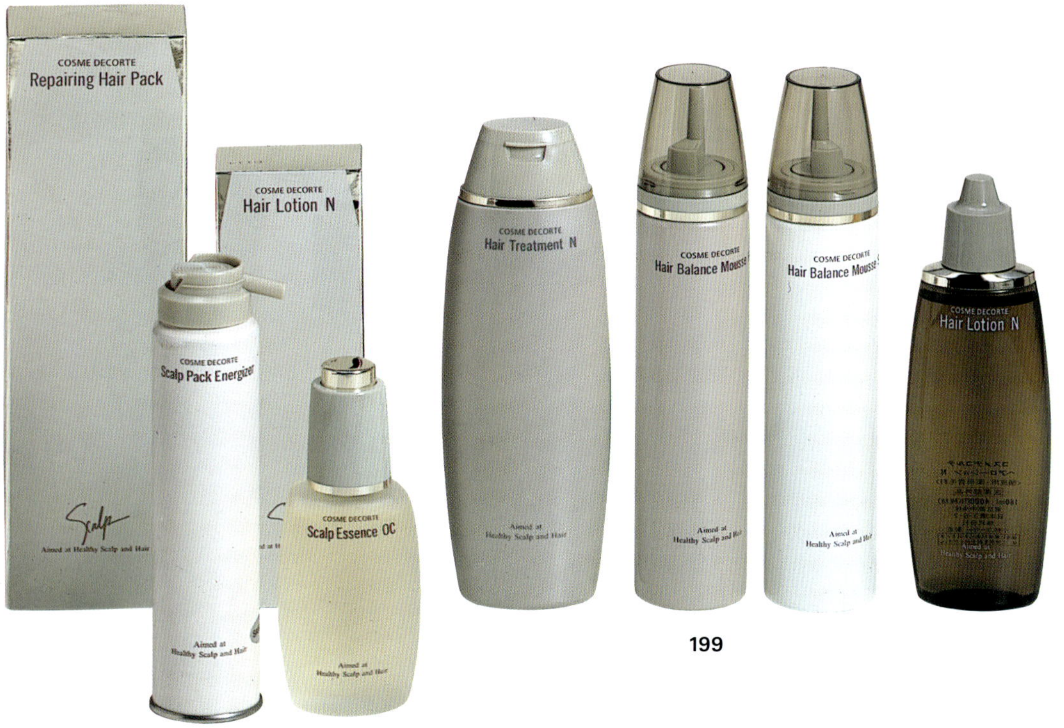

199

199 コスメデコルテ
ヘア&スカルプケアシリーズ

CD 中山紀子
AD 塙冨士雄
D 井口佳己
渡辺 均
鈴木智晴
CL コーセー

Hair-Care Products

CD Michiko Nakayama
AD Fujio Hanawa
D Yoshimi Iguchi
Hitoshi Watanabe
Chiharu Suzuki
CL Kosé

201

200

200 アルビオン エクシア

AD 錦織章三
D 黒岩光彦
倉重さおり
CL アルビオン

Cosmetics

AD Shozo Nishigori
D Mitsuhiko Kuroiwa
Saori Kurashige
CL Albion

201 ステラマリス化粧品

AD 竹林善孝
D 竹林善孝
後藤黄太
CL 日本生活協同組合連合会

Cosmetics

AD Yoshitaka Takebayashi
D Yoshitaka Takebayashi
Kota Goto
CL Japanese Consumers' Co-Operative Union

202

202 コスメデコルテ **AQ** ベースメイクシリーズ

CD　中山紀子
AD　塙冨士雄
D　井口佳己
　　矢島 修
　　渡辺 均
　　鈴木智晴
CL　コーセー

Cosmetics

CD　Michiko Nakayama
AD　Fujio Hanawa
D　Yoshimi Iguchi
　　Osamu Yajima
　　Hitoshi Watanabe
　　Chiharu Suzuki
CL　Kosé

203

203 YSスキンケアシリーズ

AD　熊井保亥
D　金子陽一
CL　資生堂

Cosmetics Foods

AD　Yasui Kumai
D　Yoichi Kaneko
CL　Shiseido

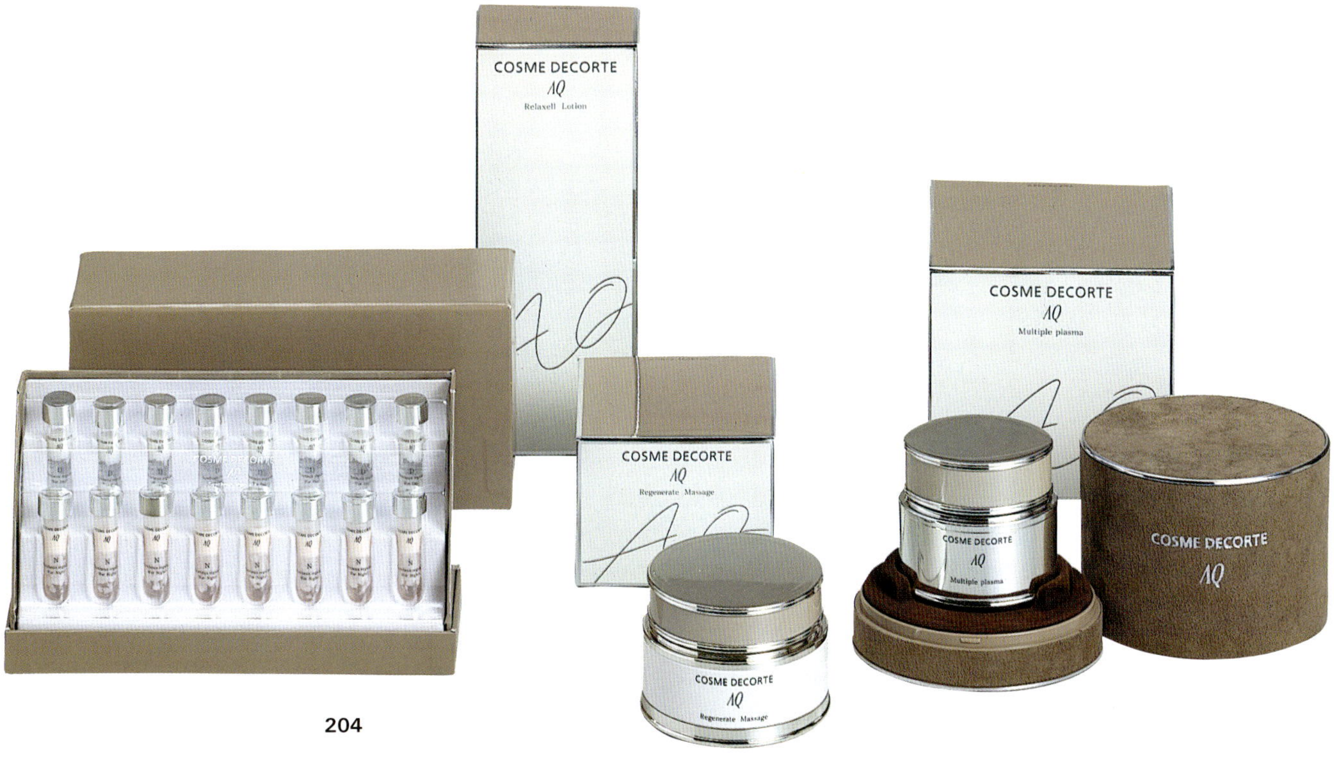

204

204 コスメデコルテ AQトリートメントシリーズ

CD　中山紀子
AD　塙冨士雄
D　林 千博
　　乗富茂美
　　石橋利江子
　　矢島 修
　　横倉尚子
CL　コーセー

Cosmetics

CD　Michiko Nakayama
AD　Fujio Hanawa
D　Chihiro Hayashi
　　Shigemi Noritomi
　　Rieko Ishibashi
　　Osamu Yajima
　　Naoko Yokokura
CL　Kosé

205

205 ユロコス オーテコロン

AD 宇田 優
D 広谷龍児
CL ラモコス

Fragrance

AD Masaru Uda
Dy Ryuji Hirotani
CL Lamocos

206

206 シーボン トリートメントコンセントレート

AD 宇塚 智
D 宇塚 智
吉本博美
谷口ひとみ
CL シーボン化粧品総合本舗

Cosmetics

AD Satoshi Uzuka
D Satoshi Uzuka
Hiromi Yoshimoto
Hitomi Taniguchi
CL C'bon Cosmetics

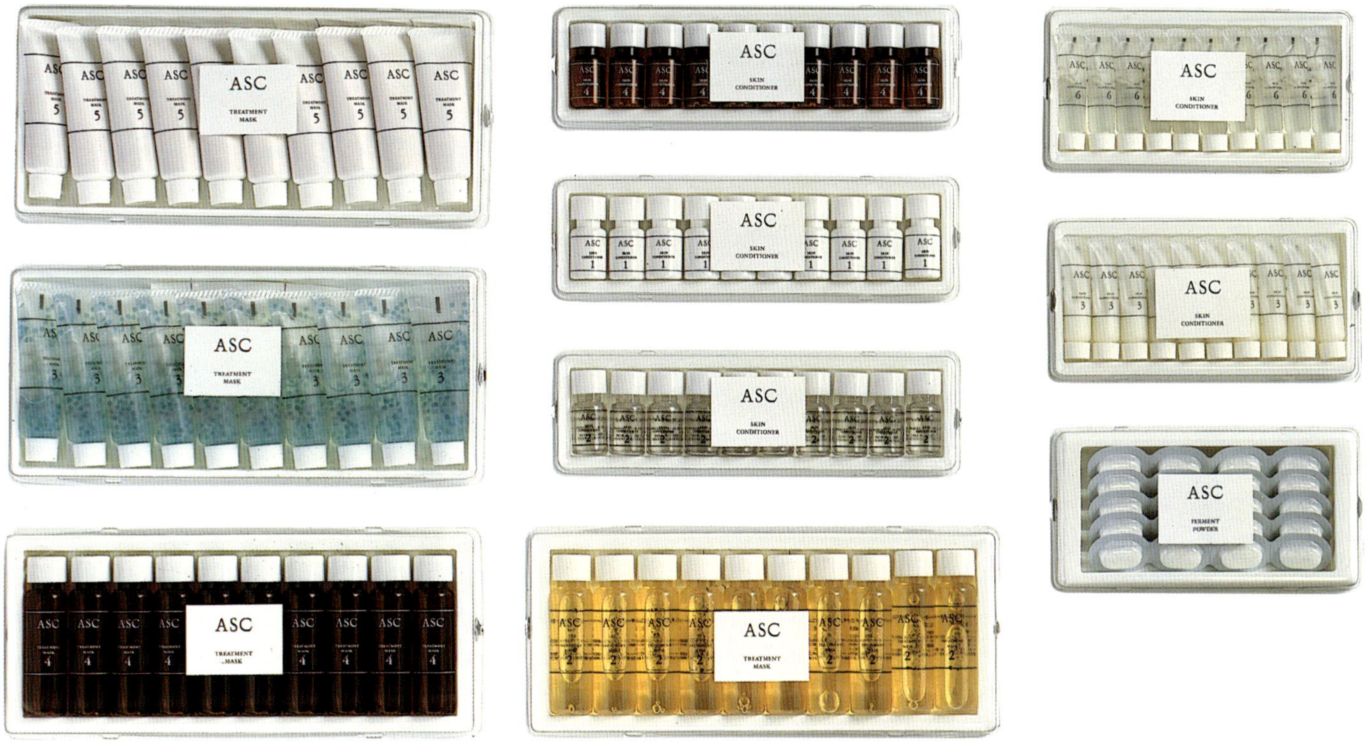

207

207 A.S.C.シリーズ
AD 宇塚 智
D 宇塚 智
吉本博美
CL シーボン化粧品総合本舗

Cosmetics
AD Satoshi Uzuka
D Satoshi Uzuka
Hiromi Yoshimoto
CL C'bon Cosmetics

208 ネクスト ファインコロンシリーズ
PL 山田博子
CD 山田博子
AD 山田博子
D バウス
CL コスメポート

Fragrance
PL Hiroko Yamada
CD Hiroko Yamada
AD Hiroko Yamada
D Baus
CL Cosmeport

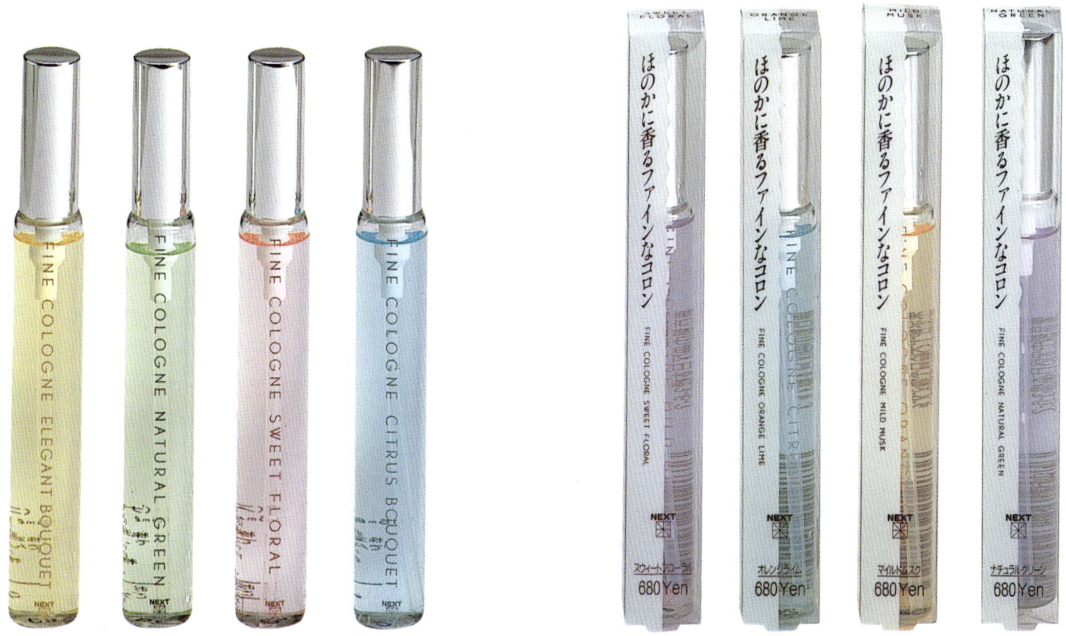

208

209

209 サンコートシリーズ

AD　人見清志
D　山本好詩
　　吉川久代
CL　ノエビア

Suntan Sunscreen Products

AD　Kiyoshi Hitomi
D　Koshi Yamamoto
　　Hisayo Yoshikawa
CL　Noevir

210 カネボウ
ウォータープルーフアイメイクアップ

AD　東海林俊郎
D　佐竹愼一郎
　佐竹ヒロコ
CL　カネボウ化粧品

Cosmetics

AD　Toshiro Shoji
D　Shin'ichiro Satake
　Hiroko Satake
CL　Kanebo Cosmetics

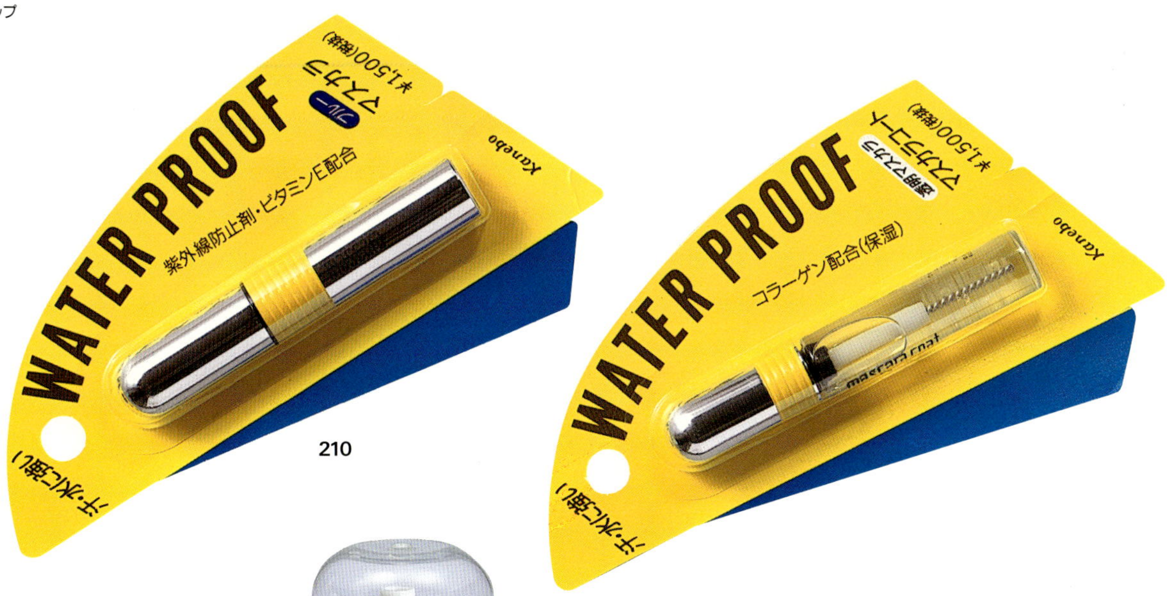

210

211 UVムースアンダーベース

CD　植村 秀
AD　中塚大輔
D　白柳悦子
CL　シュウウエムラ化粧品

Cosmetics

CD　Shu Uemura
AD　Daisuke Nakatsuka
D　Etsuko Shirayanagi
CL　Shu Uemura Cosmetics

212 リップアミュレット

AD　平戸絵里子
D　四田千絵子
IL　四田千絵子
CL　資生堂

Lip Cream

AD　Eriko Hirato
D　Chieko Yotsuda
IL　Chieko Yotsuda
CL　Shiseido

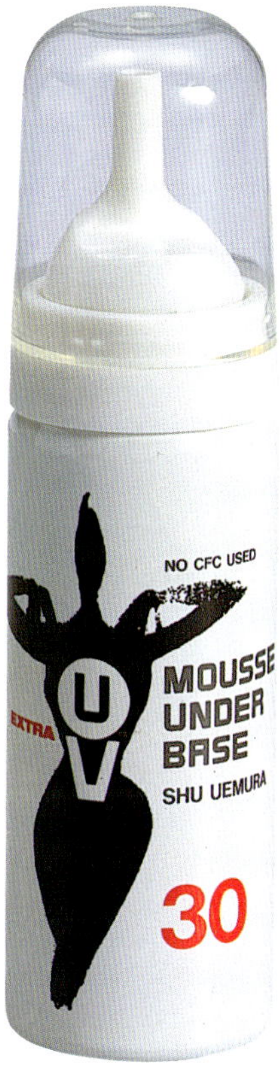

211

212

213

213 アルビオン プレゼンテーション

AD 錦織章三
D 野田徳雄
吉田なお子
CL アルビオン

Cosmetics

AD Shozo Nishigori
D Norio Noda
Naoko Yoshida
CL Albion

214

214 リシェール

CAD 太田雅雄
AD 中井一郎
D 中井一郎
高橋新三
ヘルムート・シュミット
CL 資生堂

Hair Treatment

CAD Masao Ota
AD Ichiro Nakai
D Ichiro Nakai
Shinzo Takahashi
Helmut Schmid
CL Shiseido

215 ファスティナ
AD　平井宥子
D　平井宥子
　野村るり子
　和栗正代
CL　御木本製薬

Cosmetics
AD　Yuko Hirai
D　Yuko Hirai
　Ruriko Nomura
　Masayo Waguri
CL　Mikimoto Pharmaceutical

215

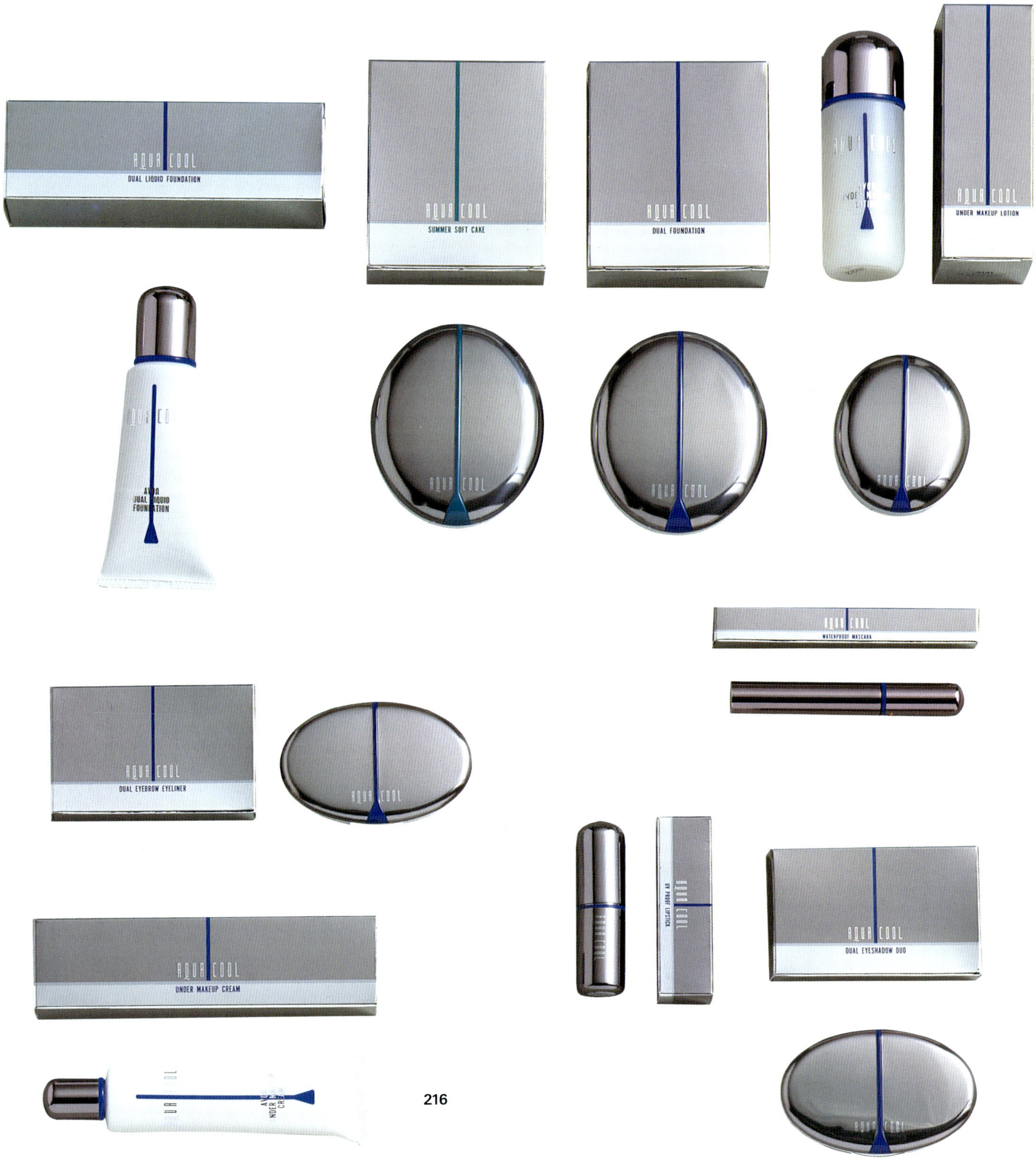

216

216 アクアクール

CD 大浜良友
AD 梅沢一徳
D 織間みさ
CL エイボン・プロダクツ

Cosmetics

CD Yoshitomo Ohama
AD Kazunori Umezawa
D Misa Orima
CL Avon Products

217

217 ポーラ アフルースメーク
AD　八野田忠
D　浪川恵美子
CL　ポーラ化粧品本舗

Cosmetics
AD　Tadashi Hachinoda
D　Emiko Namikawa
CL　Pola

218

218 カネボウ ユニフィ

PL 滝沢三良
CD 真志野みち子
AD 橋本近雄
GD 菊地 学
PD 真藤 哲
CL 鐘紡

Hair-Care Products

PL Mitsuyoshi Takizawa
CD Michiko Mashino
AD Chikao Hashimoto
GD Manabu Kikuchi
PD Akira Shindo
CL Kanebo

220 サクセスエクセレントシリーズ

AD 三井卓夫
森川洋之介
D 永井 満
渡辺隆史
CL 花王

Cosmetics

AD Takuo Mitsui
Yonosuke Morikawa
D Mitsuru Nagai
Takashi Watanabe
CL Kao

220

219

219 資生堂 化粧用具シリーズ
PL 村田智明
AD 村田智明
池田修一
D 村田智明
高橋伸幸
清水 礼
CL 資生堂

Cosmetics Goods
PL Chiaki Murata
AD Chiaki Murata
Shuichi Ikeda
D Chiaki Murata
Nobuyuki Takahashi
Aya Shimizu
CL Shiseido

221	ナチュラルヘアーソープ ウィズシリーズ		Hair-Care Products
PL	保科 高	PL	Takashi Hoshina
AD	GK京都	AD	GK Kyoto
D	GK京都	D	GK Kyoto
BD	GK京都	BD	GK Kyoto
IL	GK京都	IL	GK Kyoto
CL	タカラベルモント	CL	Takara Belmont

221

222 ルーティ
AD 滝沢三良
D 小玉章文
小国喜久子
BD 加藤貴穂
CL 鐘紡

Shampoo Conditioner
AD Mitsuyoshi Takizawa
D Akifumi Kodama
Kikuko Oguni
BD Mizuho Kato
CL Kanebo

222

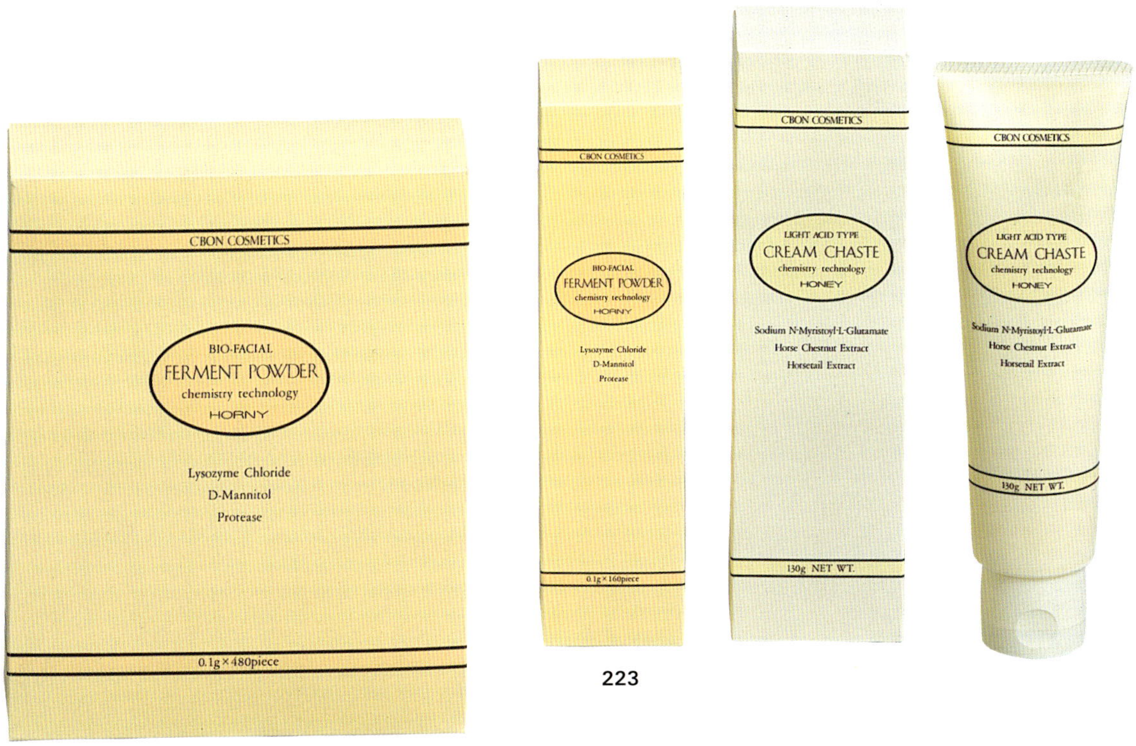

223

223 シーボン ウォッシュライン
AD 宇塚 智
D 宇塚 智
吉本博美
CL シーボン化粧品総合本舗

Cosmetics
AD Satoshi Uzuka
D Satoshi Uzuka
Hiromi Yoshimoto
CL C'bon Cosmetics

224

224 フィットフィックス
ボディーケアシリーズ

AD 池田修一
D 鐘尾 隆
飯田和子
LD 鐘尾 隆
CL 資生堂

Cosmetics

AD Shuichi Ikeda
D Takashi Kaneo
Kazuko Iida
LD Takashi Kaneo
CL Shiseido

225 アルビオン
ベジタブルウォッシュ

AD 錦織章三
D 小泉陽一
CL アルビオン

Cosmetics

AD Shozo Nishigori
D Yoichi Koizumi
CL Albion

225

226 ボディービューティ

CD 大浜良友
AD 梅沢一徳
D 村田美佳
IL 岡地裕美子
CL エイボン・プロダクツ

Bodycare Products

CD Yoshitomo Ohama
AD Kazunori Umezawa
D Mika Murata
IL Yumiko Okaji
CL Avon Products

226

227

228

227 サンジェニック

AD 山下正和
足立和彦
D 大智 恵
LO ヘッズ
CL 資生堂

Suntan Products

AD Masakazu Yamashita
Kazuhiko Adachi
D Megumu Ochi
LO Heads
CL Shiseido

228 サンダウン

AD 半田広喜
D 半田広喜
和田 徹
CL ジョンソン・エンド・ジョンソン

Sunscreen Products

AD Hiroki Handa
D Hiroki Handa
Toru Wada
CL Johnson and Johnson

229

230

229 コパトーン スポーツライン

AD トッパンパッケージングセンター
D 小宮 茂
IL 小宮 茂
CL シービック

Suntan Oil Sunscreen

AD Toppan Packaging Center
D Shigeru Komiya
IL Shigeru Komiya
CL Cbic

230 コパトーン スキートリオ

AD トッパンパッケージングセンター
D 小宮 茂
LO 小宮 茂
CL シービック

Suntan Oil/Sunscreen

AD Toppan Packaging Center
D Shigeru Komiya
LO Shigeru Komiya
CL Cbic

231

232

231 ワンデイ
AD 奥村昭夫
D 切間晴美
CL 牛乳石鹼共進社

Shampoo/Conditioner
AD Akio Okumura
D Harumi Kirima
CL Cow Brand Soap Kyoshinsha

232 泡コート/霧コート
AD 奥村昭夫
D 野上周一
CL 牛乳石鹼共進社

Hair-Care Products
AD Akio Okumura
D Shuichi Nogami
CL Cow Brand Soap Kyoshinsha

233

234

233 メンズスタイリングムース

AD 山下正和
足立和彦
LO 高橋悦夫
D 竹内祥記
CL 資生堂

Cosmetics for Men

AD Masakazu Yamashita
Kazuhiko Adachi
LO Etsuo Takahashi
D Yoshiki Takeuchi
CL Shiseido

234 プロセスボディーシャンプー

AD 山田 満
D 長尾敦子
CL ツムラ

Soap

AD Mitsuru Yamada
D Atsuko Nagao
CL Tsumura

235

235 ボディーシャンプーS&C、ボディーミルクS&P

AD 村瀬省三
D 吉田 純
内田元和
村瀬 潤
CL マリーブ化粧品

Bodycare Products

AD Shozo Murase
D Jun Yoshida
Motokazu Uchida
Jun Murase
CL Maleave Cosmetics

236

236 クールプロセスボディーシャンプー

AD 山田 満
D 長尾敦子
IL 飯田 淳
CL ツムラ

Soap

AD Mitsuru Yamada
D Atsuko Nagao
IL Jun Iida
CL Tsumura

237

237 アルビオン ボディーコントロール

AD　錦織章三
D　吉田なお子
CL　アルビオン

Bodycare Products

AD　Shozo Nishigori
D　Naoko Yoshida
CL　Albion

238 ILZAX

CD 藤元 靖
AD 竹智 淳
D 竹智 淳
CL 東京ビューティーセンター

Shampoo

CD Yasushi Fujimoto
AD Jun Takechi
D Jun Takechi
CL Tokyo Beauty Center

239 ソフトスタイル

AD 花村賢二
D 重田元恵
CL マンダム

Hair-Care Products

AD Kenji Hanamura
D Motoe Shigeta
CL Mandom

240

240 アルビオン ディセ

AD 錦織章三
小泉陽一
D 中野 恵
CL アルビオン

Hair-Care Products

AD Shozo Nishigori
Yoichi Koizumi
D Megumi Nakano
CL Albion

241 クアタイム

CD　木村 滋
AD　博報堂
D　博報堂
CL　エーザイ

Bath Products

CD　Shigeru Kimura
AD　Hakuhodo
D　Hakuhodo
CL　Eisai

242 モイスチャーミルク・バブ、ミルキーフレッシュ・バブ

AD　吉田光利
D　平澤雅子
CL　花王

Bath Products

AD　Mitsutoshi Yoshida
D　Masako Hirasawa
CL　Kao

241

242

243

243 四季の香り

AD　田中康夫
GD　田中康夫
IL　野田 徹
CL　クローバーコーポレーション
伊藤忠ファッションシステム

Bath Products

AD　Yasuo Tanaka
GD　Yasuo Tanaka
IL　Toru Noda
CL　Clover
Ito-Chu Fashion System

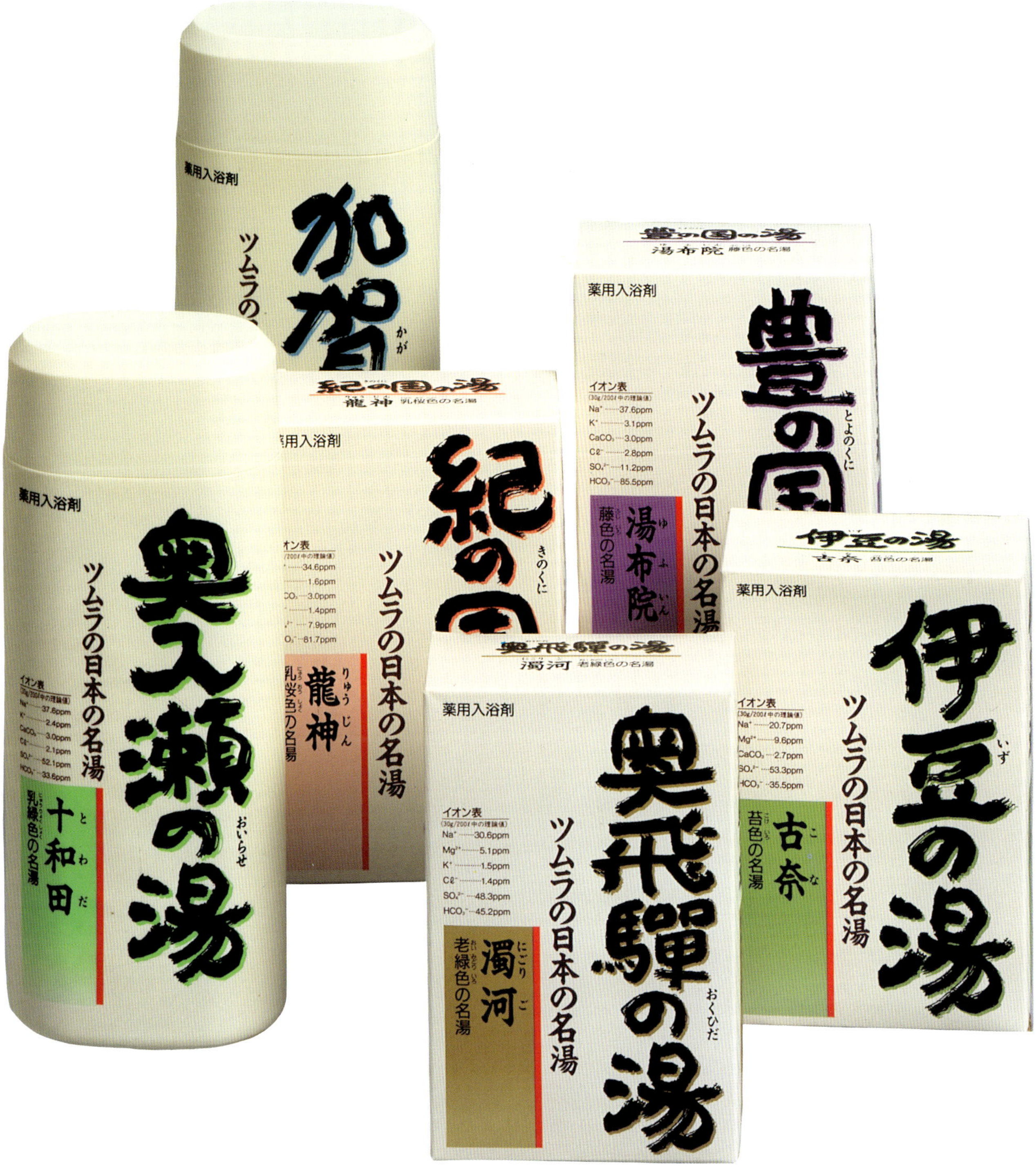

244

244 日本の名湯シリーズ

AD　山田 満
D　川路ヨウセイ
　宮本泰志
CL　ツムラ

Bath Products

AD　Mitsuru Yamada
D　Yosei Kawaji
　Yasushi Miyamoto
CL　Tsumura

一般雑貨 家庭用品
医薬品 衣料品

Varieties/Household Commodities/Medicines/Clothes

245

245 センシャス
AD　ハーズ実験デザイン研究所
　　村田智明
D　村田智明
　　三浦康男
CL　ナワタ

Toy
AD　Hers Experimental Design Laboratory
　　Chiaki Murata
D　Chiaki Murata
　　Yasuo Miura
CL　Nawata

246

247

246 エコロジーカードケース

AD 狩野浩之
D 狩野浩之
Const 狩野浩之
杉山進一

Visiting Card Case

AD Hiroyuki Kano
D Hiroyuki Kano
Const Hiroyuki Kano
Shin'ichi Sugiyama

247 あかまるきいろさんかく

AD ハーズ実験デザイン研究所
村田智明
D 村田智明
三浦康男
CL ナワタ

Toy

AD Hers Experimental Design Laboratory
Chiaki Murata
D Chiaki Murata
Yasuo Miura
CL Nawata

248 harf

AD 藤井陽一郎
P 野寺治孝
CL Slow Hand

Calendar

AD Yoichiro Fujii
P Harutaka Nodera
CL Slow Hand

248

249 キューブロック

PL 藤島武士
AD 藤島武士
D 藤島武士
GD 藤島武士
PD 藤島武士
PR 東 一雄(日本リツビ)
CL ペンギンワックス

Deodorant

PL Takeshi Fujishima
AD Takeshi Fujishima
D Takeshi Fujishima
GD Takeshi Fujishima
PD Takeshi Fujishima
PR Kazuo Azuma(Nihon Ritsubi)
CL Penguin Wax

249

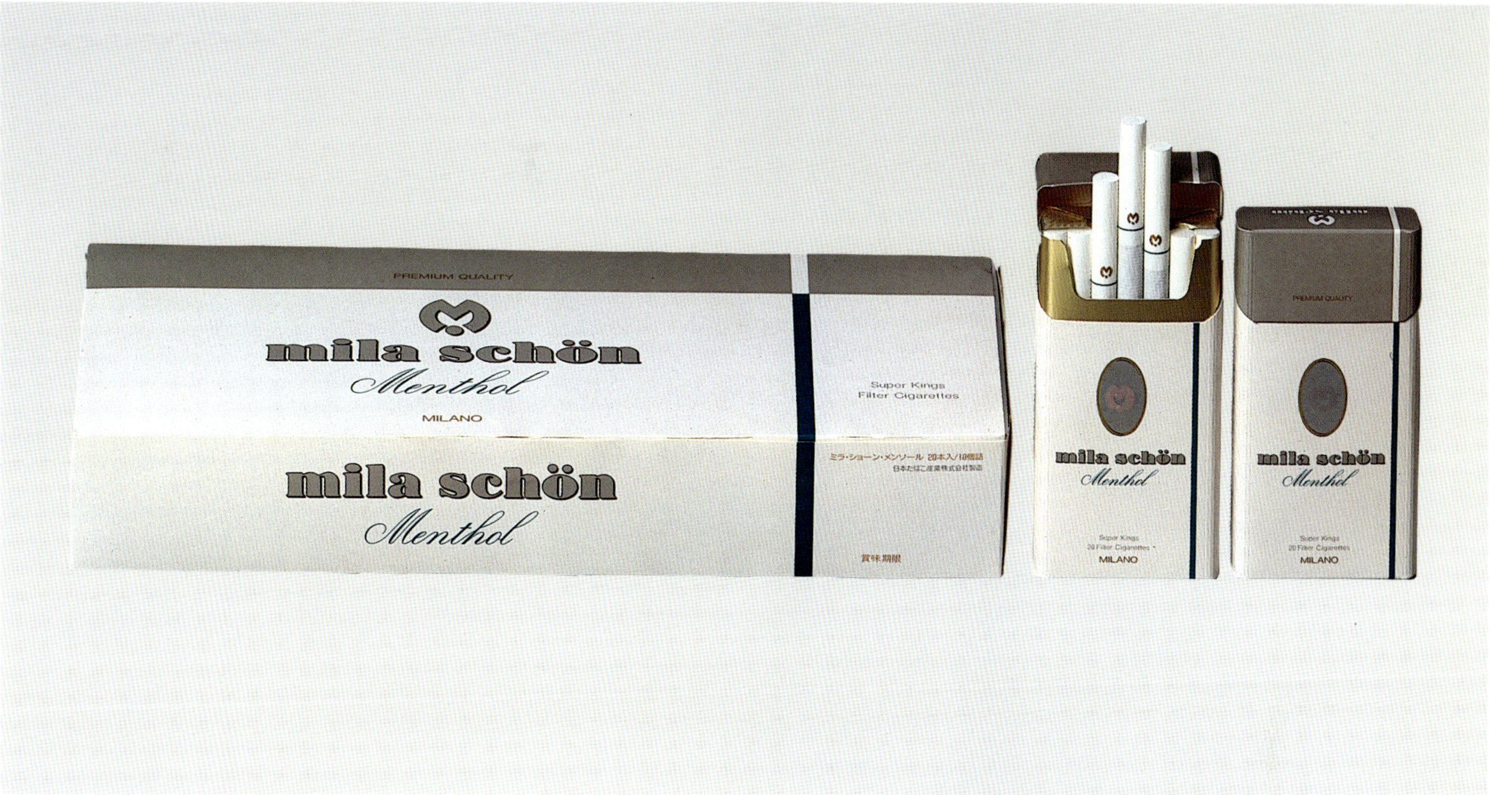

250

250 ミラーショーンシガレット

PL　垣内貞夫
AD　垣内貞夫
D　垣内貞夫
CL　日本たばこ産業

Cigarettes

PL　Sadao Kakiuchi
AD　Sadao Kakiuchi
D　Sadao Kakiuchi
CL　Japan Tabacco

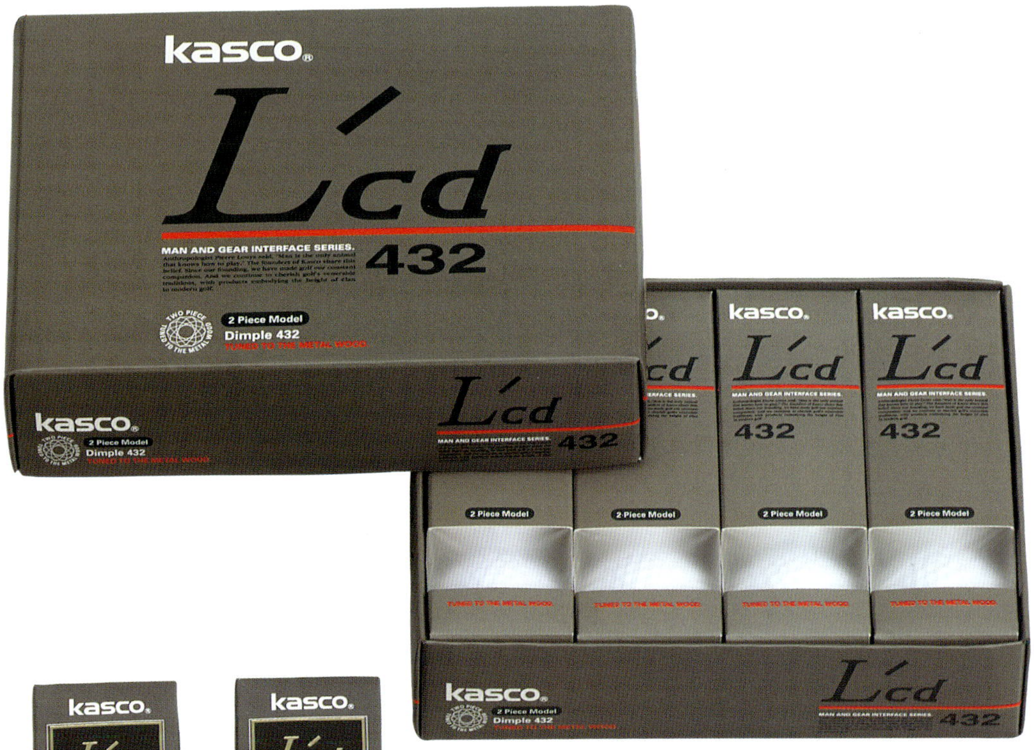

251

251 キャスコ ゴルフボールL'cd

CD　生西正幸
AD　河村岳志
D　中辻智大
　　河村岳志
　　FDグラフィック
CL　キャスコ

Golf Ball

CD　Masayuki Ikunishi
AD　Takeshi Kawamura
D　Tomohiro Nakatsuji
　　Takeshi Kawamura
　　FD Graphic
CL　Kasco

252

252 ダイキン レザノヴァ
ゴルフグローブ

AD　河村岳志
D　FDグラフィック
CL　ダイキン

Golf Glove

AD　Takeshi Kawamura
D　FD Graphic
CL　Daikin

253 クララソフトコンタクトレンズ

AD　田中康夫
GD　田中康夫
CL　クラレ

Contact Lens

AD　Yasuo Tanaka
GD　Yasuo Tanaka
CL　Kuraly

253

254 シンテジフレーム

AD　佐藤 淳
D　岩 賢俊
　ミケーレ・デ・ルッキ
Const　七福商事
CL　ギャラリーインターフォーム

Photo Frame

AD　Jun Sato
D　Kenji Iwa
　Michele De Lucchi
Const　Shichifuku Shoji
CL　Gallery Interform

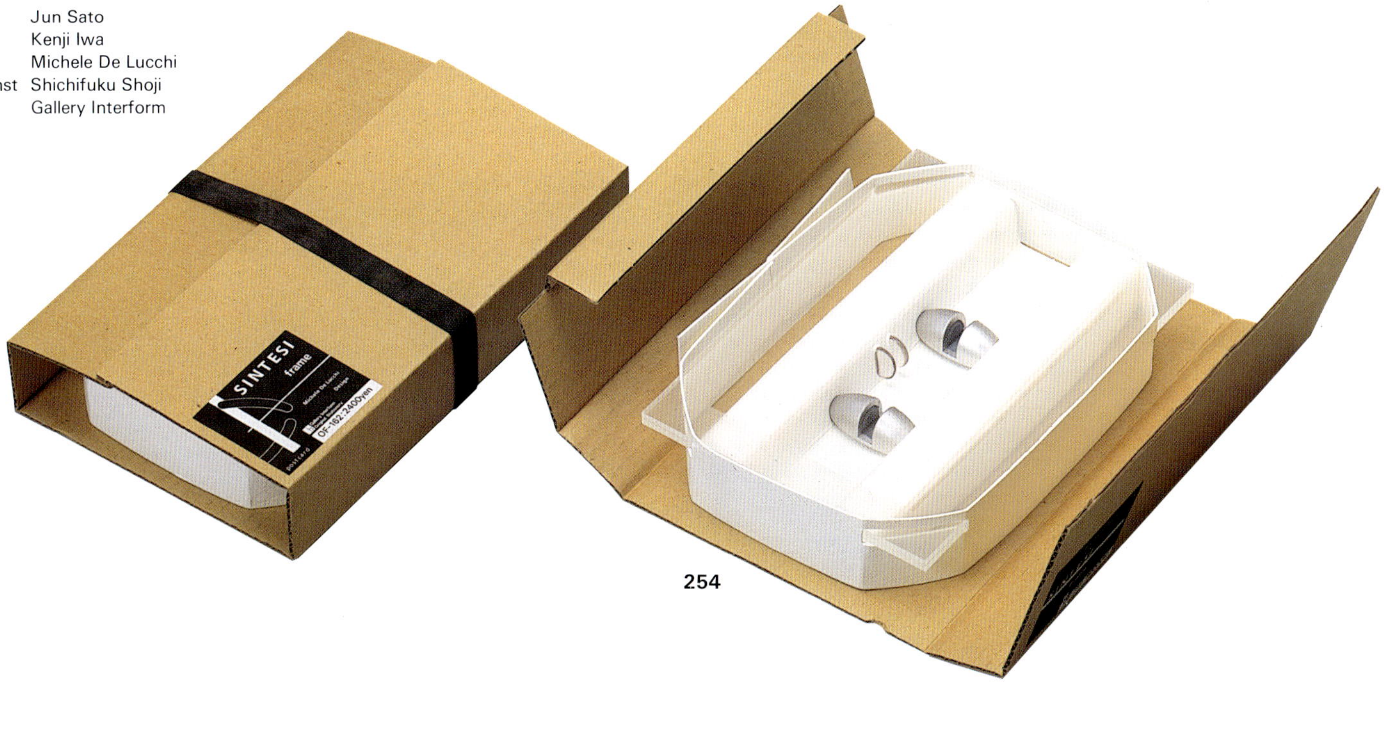

254

255 クリネックス
スーパードライタオル

AD 小幡義晴
D 岡田宏三
田中春雄
CL 十條キンバリー

Kitchen Towels

AD Yoshiharu Obata
D Kożo Okada
Haruo Tanaka
CL Jujo Kimberly

255

256 クリネックス ポケットティッシュ
AD 小幡義晴
D 岡田宏三
田中春雄
CL 十條キンバリー

Tissue
AD Yoshiharu Obata
D Kozo Okada
Haruo Tanaka
CL Jujo Kimberly

257 クリネックス キッチンタオル
AD 小幡義晴
D 岡田宏三
田中春雄
CL 十條キンバリー

Kitchen Towels
AD Yoshiharu Obata
D Kozo Okada
Haruo Tanaka
CL Jujo Kimberly

256

257

258 クリネックス ブティーク
AD 小幡義晴
D 川路ヨウセイ
近藤和磨
CL 十條キンバリー

Tissue
AD Yoshiharu Obata
D Yosei Kawaji
Kazuma Kondo
CL Jujo Kimberly

258

259 ルックシリーズ

CD　岡田一朗
AD　野崎靖紀
　　藤田次雄
D　福田環境デザイン研究所
　　デザイン・デルタ
CL　ライオン

Household Cleanser

CD　Ichiro Okada
AD　Yasutoshi Nozaki
　　Tsugio Fujita
D　Fukuda Environment Design
　　Design Delta
CL　Lion

259

260

260 ダスキン 台所用洗剤シリーズ

AD 犬塚達美
D 犬塚達美
BD 犬塚達美
近藤純夫
CL ダスキン

Dish Washing Detergent

AD Tatsumi Inuzuka
D Tatsumi Inuzuka
BD Tatsumi Inuzuka
Sumio Kondo
CL Duskin

261

262

261 タッチ1/3

AD 吉田光利
D 平澤雅子
湯本免郎
IL 木村桂子
CL 花王

Fabric Softener

AD Mitsutoshi Yoshida
D Masako Hirasawa
Itsuro Yumoto
IL Keiko Kimura
CL Kao

262 ハミング バックインボックス

AD 吉田光利
D 平澤雅子
青山明輝
CL 花王

Fabric Softener

AD Mitsutoshi Yoshida
D Masako Hirasawa
Akiteru Aoyama
CL Kao

263

264

263 ピチット

PL　那波多目康子
AD　那波多目康子
D　那波多目デザイン事務所
　　佐藤邦子
LO　佐藤邦子
P　佐藤洋一
ST　家中あい
CL　昭和電工

Absorbent Cooking Paper

PL　Koko Nabatame
AD　Koko Nabatame
D　Nabatame Design Office
　　Kuniko Sato
LO　Kuniko Sato
P　Yoichi Sato
ST　Ai Ienaka
CL　Showa Denko

264 チャーミーグリーン

CD　岡田一朗
AD　富樫泉吉
D　デザイン・マック
　　デザイン・デルタ
CL　ライオン

Dish Washing Detergent

CD　Ichiro Okada
AD　Senkichi Togashi
D　Design Mac
　　Design Delta
CL　Lion

265

266

265 ゲインズ・キャッティ まぐろ風味

AD 松永 真
D 松永 真
鳩貝浩司
CL 味の素ゼネラルフーヅ

Pet Food

AD Shin Matsunaga
D Shin Matsunaga
Koji Hatogai
CL Ajinomoto General Foods

266 ゲインズ・ミールファインミックス

AD 松永 真
D 松永 真
鳩貝浩司
CL 味の素ゼネラルフーヅ

Pet Food

AD Shin Matsunaga
D Shin Matsunaga
Koji Hatogai
CL Ajinomoto General Foods

267 ミルパックス トライアングル

AD 高田修自
D 高田修自
Const 高田修自
CL パックス

Dustbin

AD Shuji Takada
D Shuji Takada
Const Shuji Takada
CL Pax

267

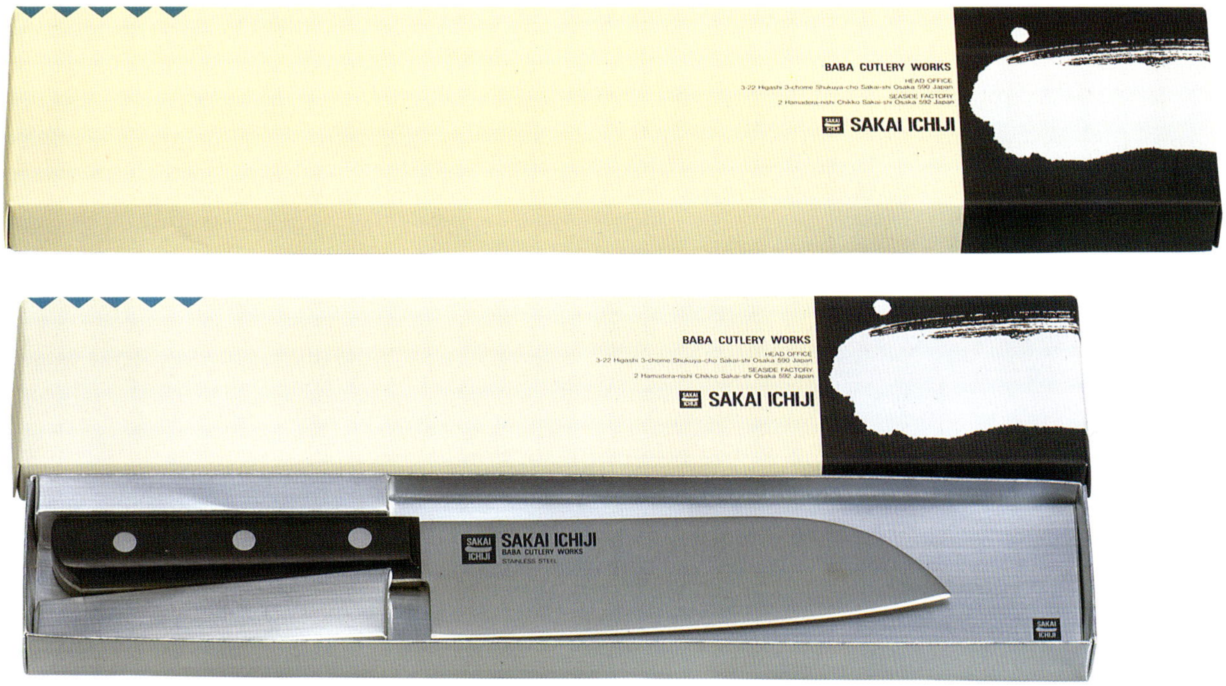

268

268 包丁パッケージ

AD 前田和紀
D 前田和紀
早川文和
LO 前田誠一
CL 馬場刃物製作所

Kitchen Knife

AD Kazuki Maeda
D Kazuki Maeda
Fumikazu Hayakawa
LO Seiichi Maeda
CL Baba Cutlery Works

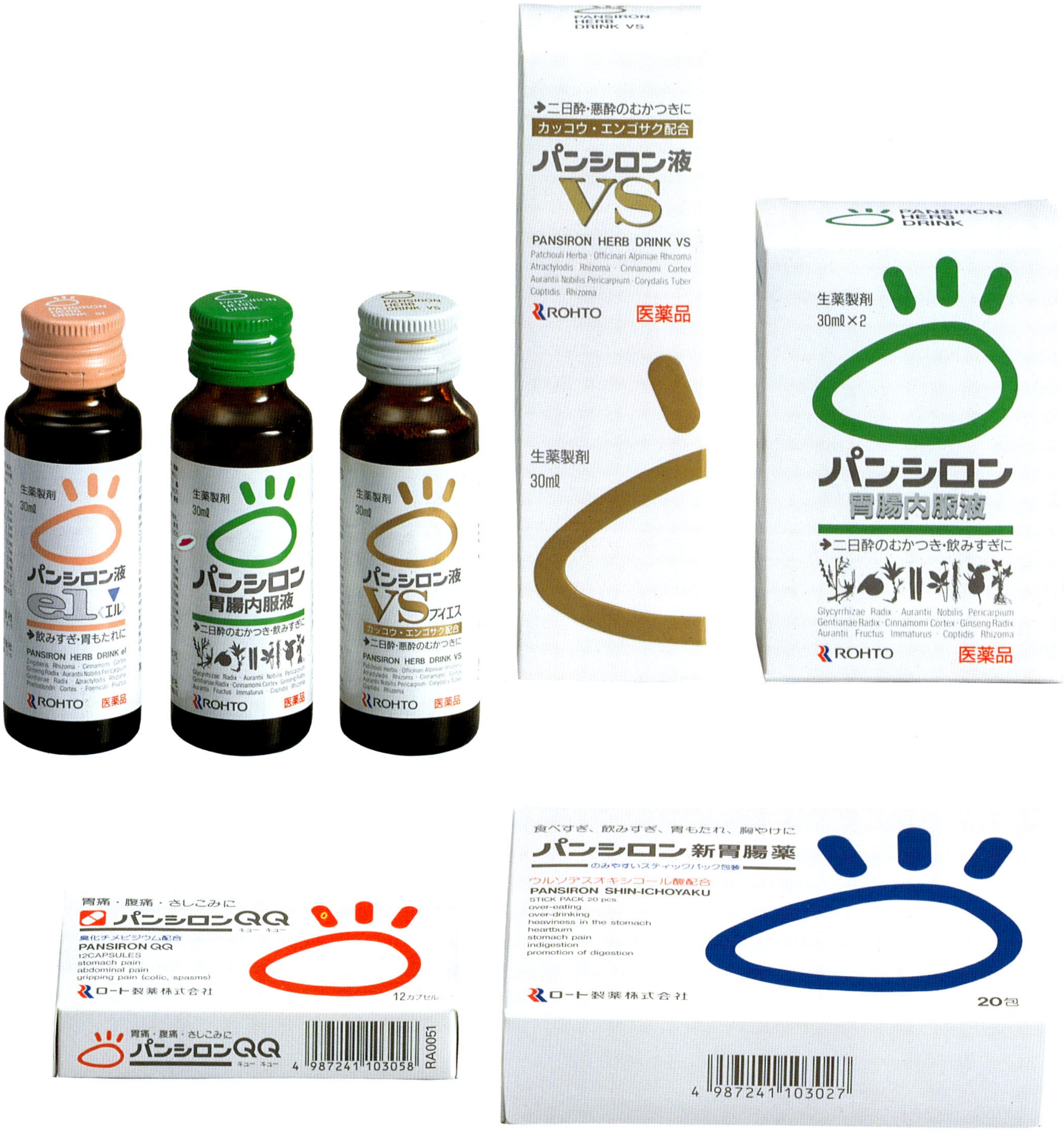

269

269 パンシロン 胃腸薬シリーズ

AD ハーズ実験デザイン研究所
村田智明
D 村田智明
高橋伸幸
清水 礼
CL ロート製薬

Medicine

AD Hers Experimental Design Laboratory
Chiaki Murata
D Chiaki Murata
Nobuyuki Takahashi
Aya Shimizu
CL Rohto Pharmaceutical

270 ライフィックスA2

CD 和田 亨
D 阿部 優
中田淑子
CL ライフィックス

Medicine

CD Toru Wada
D Masaru Abe
Yoshiko Nakada
CL Lifix

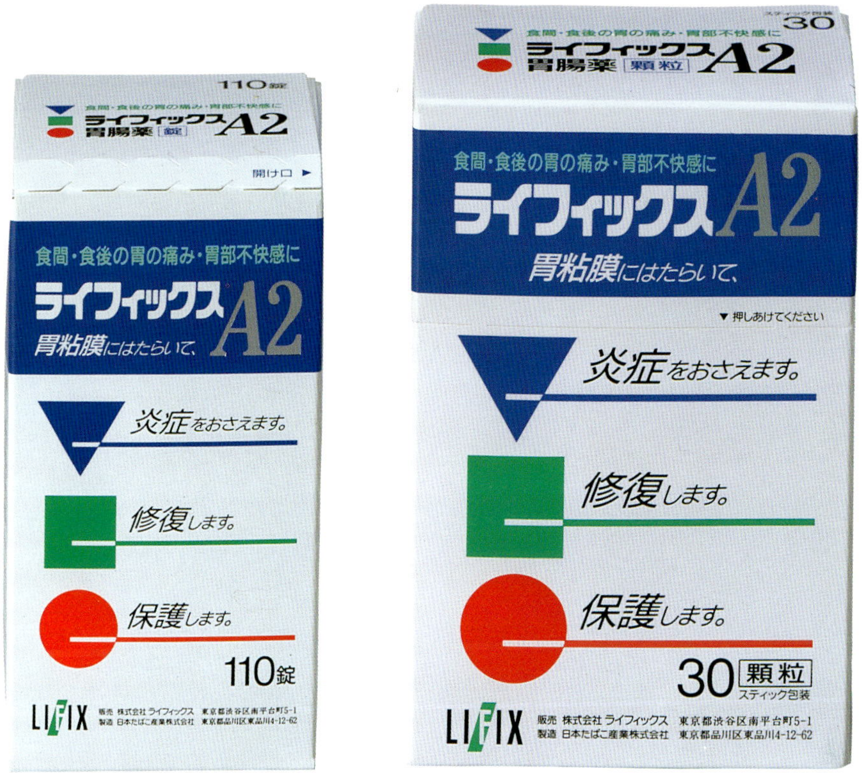

270

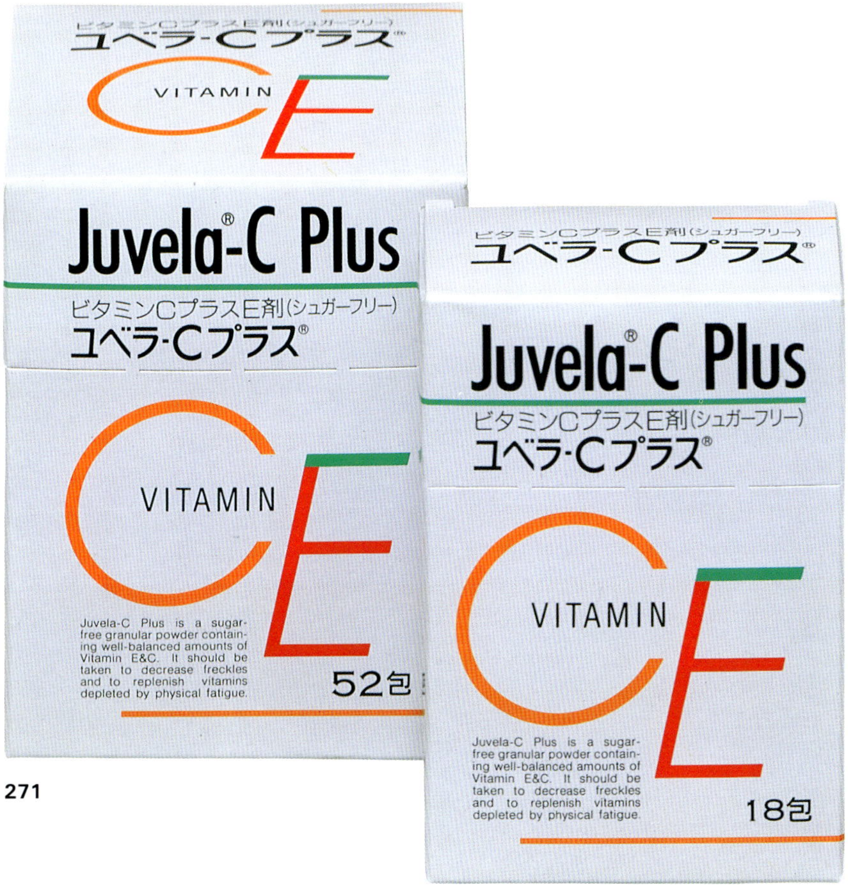

271

271 ユベラCプラス

CD 木村 滋
AD 電通
D 電通
CL エーザイ

Dietary Supplement

CD Shigeru Kimura
AD Dentsu
D Dentsu
CL Eisai

272

273

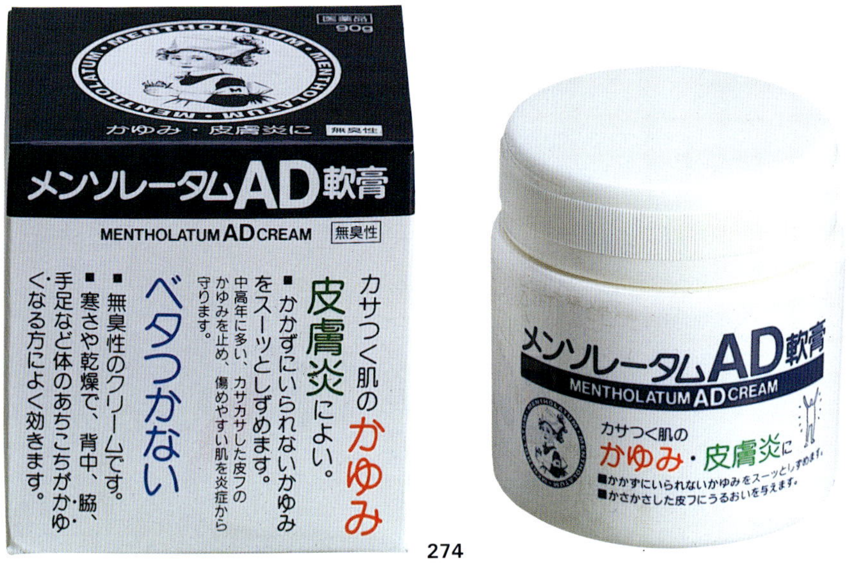

274

272 ニュートクホンチールA

PL 鈴木勇蔵
CD 宮村政男
AD 川田雅直
CL トクホン

Medical Plasters

PL Yuzo Suzuki
CD Masao Miyamura
AD Masanao Kawata
CL Tokuhon

273 ジョガーズ

AD 木全時彦
穂積雅明
D 半田広喜
福島麻里
徳本芳美
CL 佐藤製薬

Dietary Supplement

AD Tokihiko Kimata
Masaaki Hozumi
D Hiroki Handa
Mari Fukushima
Yoshimi Tokumoto
CL Sato Pharmaceutical

274 メンソレータムAD軟膏

AD ハーズ実験デザイン研究所
村田智明
D 村田智明
松尾義之
高橋伸幸
CL ロート製薬

Ointment

AD Hers Experimental Design Laboratory
Chiaki Murata
D Chiaki Murata
Yoshiyuki Matsuo
Nobuyuki Takahashi
CL Rohto Pharmaceutical

275

275 活

AD　斉藤芳弘
BD　佐藤 卓
LD　斉藤芳弘
CL　ツムラ

Dietary Supplement

AD　Yoshihiro Saito
BD　Taku Sato
LD　Yoshihiro Saito
CL　Tsumura

276

276 ノアールシリーズ
AD 木全時彦
穂積雅明
D 川路ヨウセイ
近藤和磨
CL 佐藤製薬

Medicine
AD Tokihiko Kimata
Masaaki Hozumi
D Yosei Kawaji
Kazuma Kondo
CL Sato Pharmaceutical

277 エスカルチン
CD 和久津康弘
AD 須田 哲
D 須田 哲
CL 小林製薬

Medicine
CD Yasuhiro Wakutsu
AD Satoshi Suda
D Satoshi Suda
CL Kobayashi Pharmaceutical

277

278 ボラギノール

AD　奥村昭夫
D　切間晴美
CL　天藤製薬

Medicine

AD　Akio Okumura
D　Harumi Kirima
CL　Amato Pharmaceutical Products

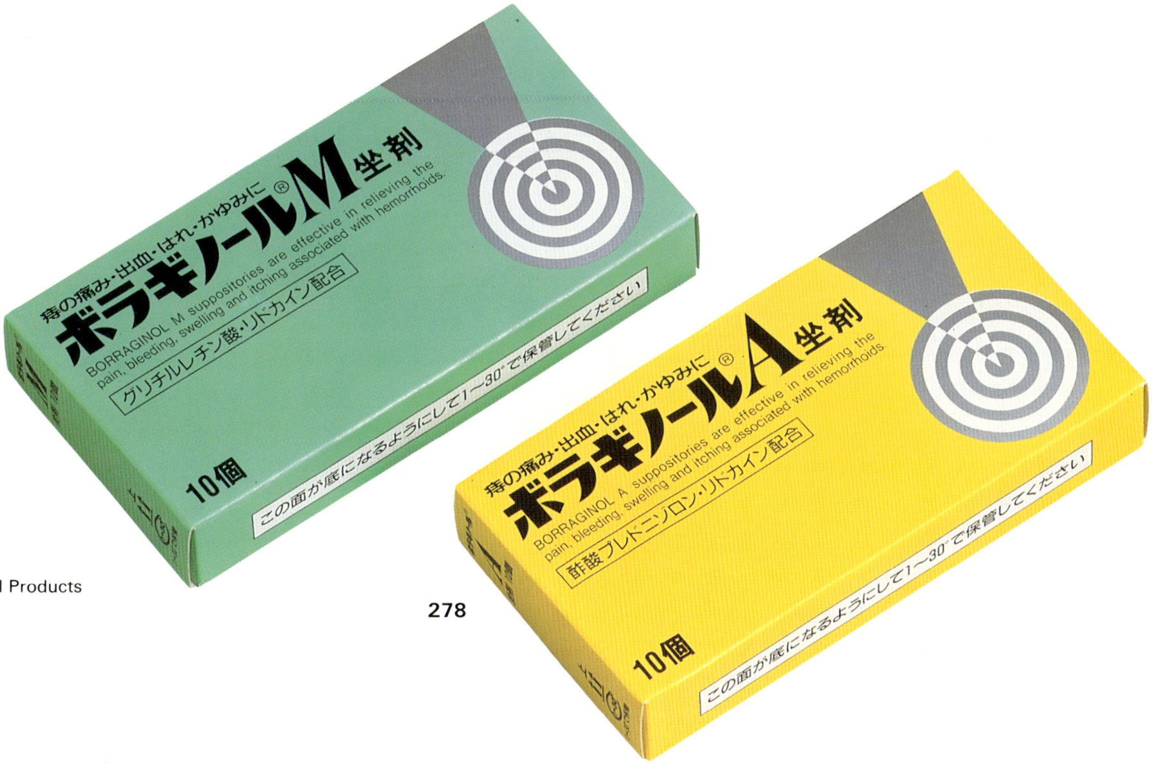

278

279 ユベラックス300

CD　木村 滋
AD　ブレーン
D　ブレーン
CL　エーザイ

Dietary Supplement

CD　Shigeru Kimura
AD　Brain
D　Brain
CL　Eisai

279

280

280 カジュアルインナー
AD 田口 圭
D 田口 圭
CL 千趣会

Clothes
AD Kei Taguchi
D Kei Taguchi
CL Senshukai

281

281 ボンヌナチュール
AD 田口 圭
D 田口 圭
CL 千趣会

Clothes
AD Kei Taguchi
D Kei Taguchi
CL Senshukai

282

282 サンエオリジン
パンティストッキング

AD　植田夏雄
GD　小柳正明
LO　小柳正明
P　宍戸真一郎
CL　サンエオリジン

Panty Hose

AD　Natsuo Ueda
GD　Masaaki Koyanagi
LO　Masaaki Koyanagi
P　Shin'ichiro Shishido
CL　San'e Origin

283

283 ホリデーシップシリーズ

AD　池田 毅
D　古久保剛
CL　千趣会

Socks

AD　Takeshi Ikeda
D　Takeshi Kokubo
CL　Senshukai

電気機器・関連商品
包装紙・ショッピングバッグ・ギフトボックス
贈答品　輸送用ケース　その他

Electric Appliances, Supplies
Wrapping Paper, Shopping Bags, Gift Boxes
Gifts/Shopping Containers/Others

284

285

284 TDK SA, MA, SF, SA-X

CD 植村 徳
AD 高橋 敏
D 小野達也
CL TDK

Cassette Tape

CD Noboru Uemura
AD Ving Takahashi
D Tatsuya Ono
CL TDK

285 TDK AD-X, SR-X

CD 植村 徳
AD 高橋 敏
D 長沢 学
CL TDK

Cassette Tape

CD Noboru Uemura
AD Ving Takahashi
D Gaku Nagasawa
CL TDK

286

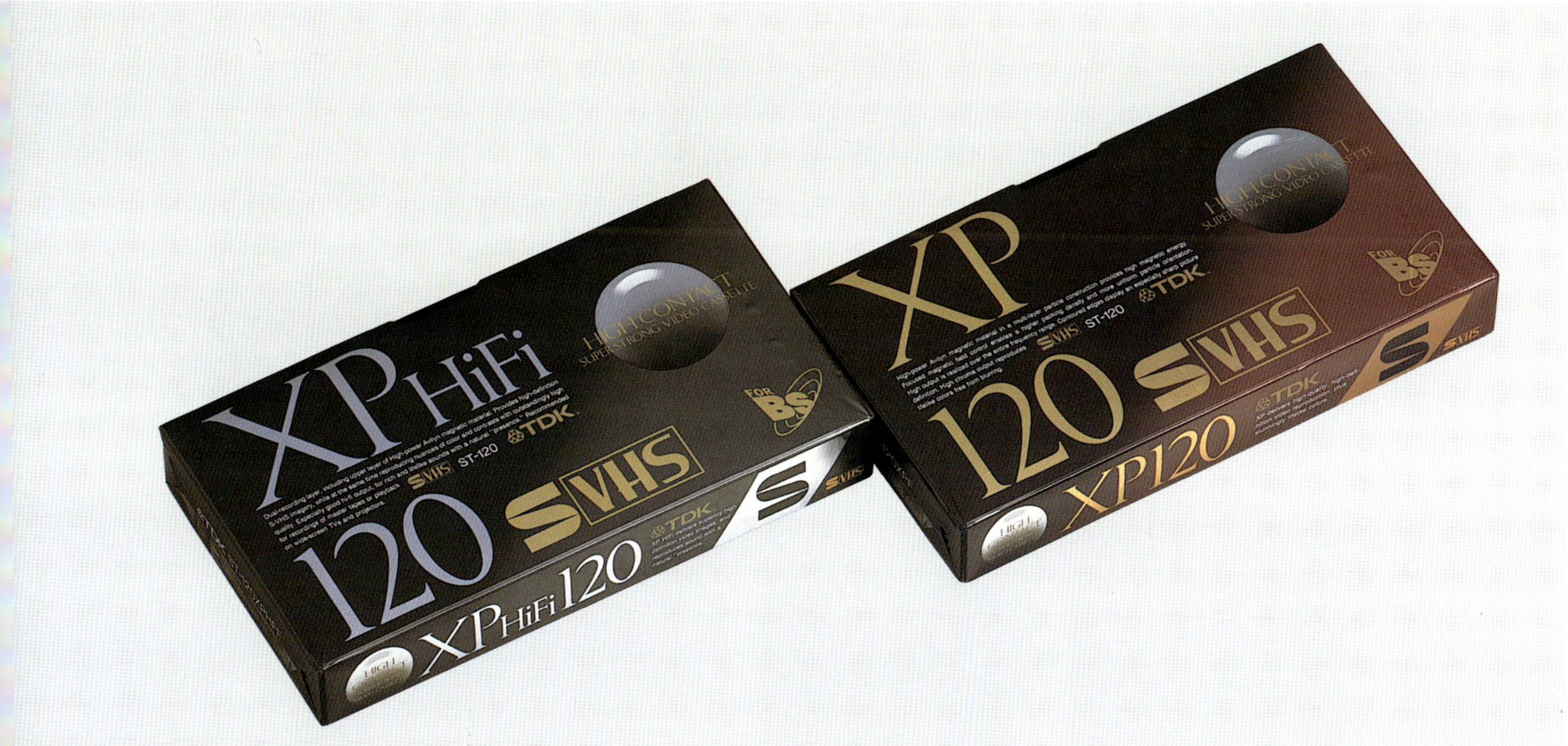

287

286 **TDK AD, SR**

CD　植村 徳
AD　高橋 敏
D　長沢 学
CL　TDK

Cassette Tape

CD　Noboru Uemura
AD　Ving Takahashi
D　Gaku Nagasawa
CL　TDK

287 **TDK XP, XP-HiFi, XP-Digital**

CD　植村 徳
AD　高橋 敏
D　長沢 学
　　高橋 敏
CL　TDK

Video Tape

CD　Noboru Uemura
AD　Ving Takahashi
D　Gaku Nagasawa
　　Ving Takahashi
CL　TDK

288

288 ビクタービデオカセット

AD 日本ビクターデザインセンター
D 日本ビクターデザインセンター
グラフィックデザイン室
CL 日本ビクター

Video Cassette

AD Design Center, Victor Company of Japan
D Graphic Design Dept., Design Center, Victor Company of Japan
CL Victor Company of Japan

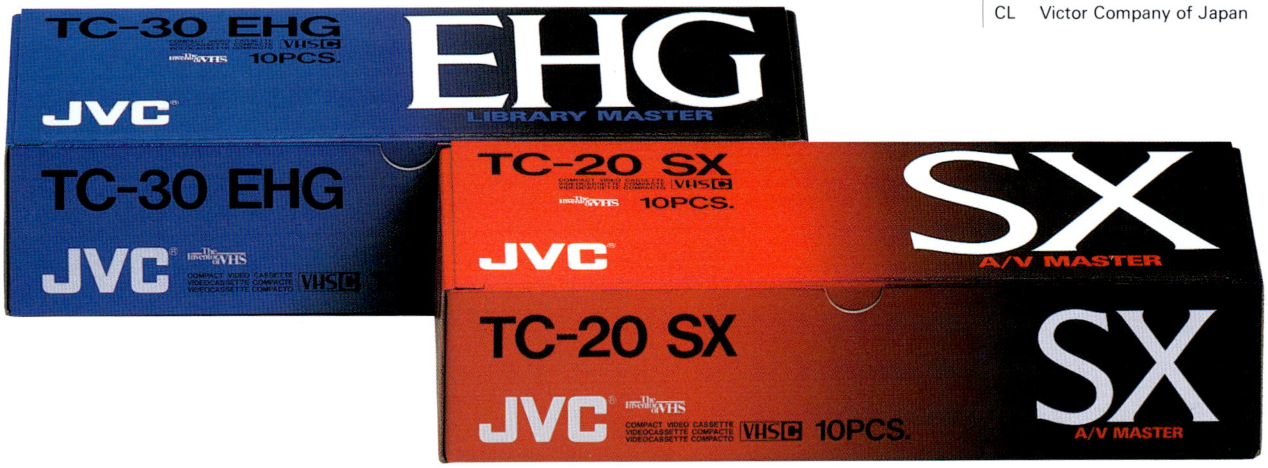

289

289 JVC ビデオカセット

AD 日本ビクターデザインセンター
D 日本ビクターデザインセンター
グラフィックデザイン室
CL 日本ビクター

Video Cassette

AD Design Center, Victor Company of Japan
D Graphic Design Dept., Design Center, Victor Company of Japan
CL Victor Company of Japan

290 三菱カラーフィルム

AD　柴﨑恵美子
D　柴﨑恵美子
CL　三菱製紙

Color Film

AD　Emiko Shibasaki
D　Emiko Shibasaki
CL　Mitsubishi Paper Mills

290

291

291 撮りっきりコニカMini
撮りっきりコニカMiniピッカリ

PL　二瓶一裕
AD　林美由紀
D　林美由紀
IL　大木裕子
CL　コニカ

Single-Use Camera

PL　Kazuhiro Nihei
AD　Miyuki Hayashi
D　Miyuki Hayashi
IL　Hiroko Oki
CL　Konica

292

292 New Music Super Box 100

AD 藤井陽一郎
D 前川恵子
P 太田マサオ
CL アルファムーン

Compact Disk

AD Yoichiro Fujii
D Keiko Maekawa
P Masao Ota
CL Alfa Moon

293

293 Hello
AD 竹智 淳
D 竹智 淳
CL キングレコード

Compact Disc
AD Jun Takechi
D Jun Takechi
CL King Record

294

294 Heart Side
AD 竹智 淳
D 竹智 淳
Copy 後藤 学
CL キングレコード

Compact Disc
AD Jun Takechi
D Jun Takechi
Copy Manabu Goto
CL King Record

295

295 東芝マイクロフロッピーディスク

CD 平井文夫
AD 真崎太郎
D 小川洋子
CL 東芝

Floppy Disk

CD Fumio Hirai
AD Taro Masaki
D Yoko Ogawa
CL Toshiba

296 東芝AVアクセサリー

CD　平井文夫
AD　真崎太郎
D　五十嵐照雄
CL　東芝

Audio Visual Accessories

CD　Fumio Hirai
AD　Taro Masaki
D　Teruo Igarashi
CL　Toshiba

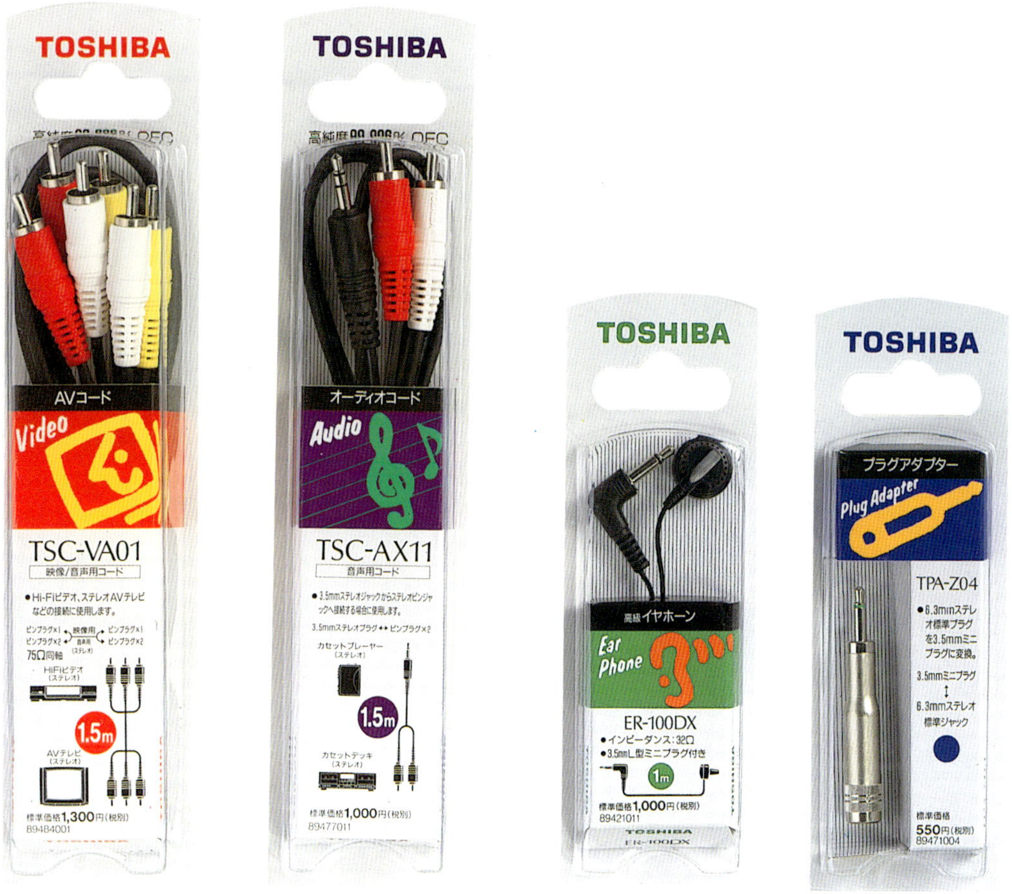

296

297

297 新しいフロッピーディスクのパッケージ提案

D　シャープ情報システムデザインセンター
CL　シャープ

Floppy Disk

D　Information Systems Design Center, Sharp
CL　Sharp

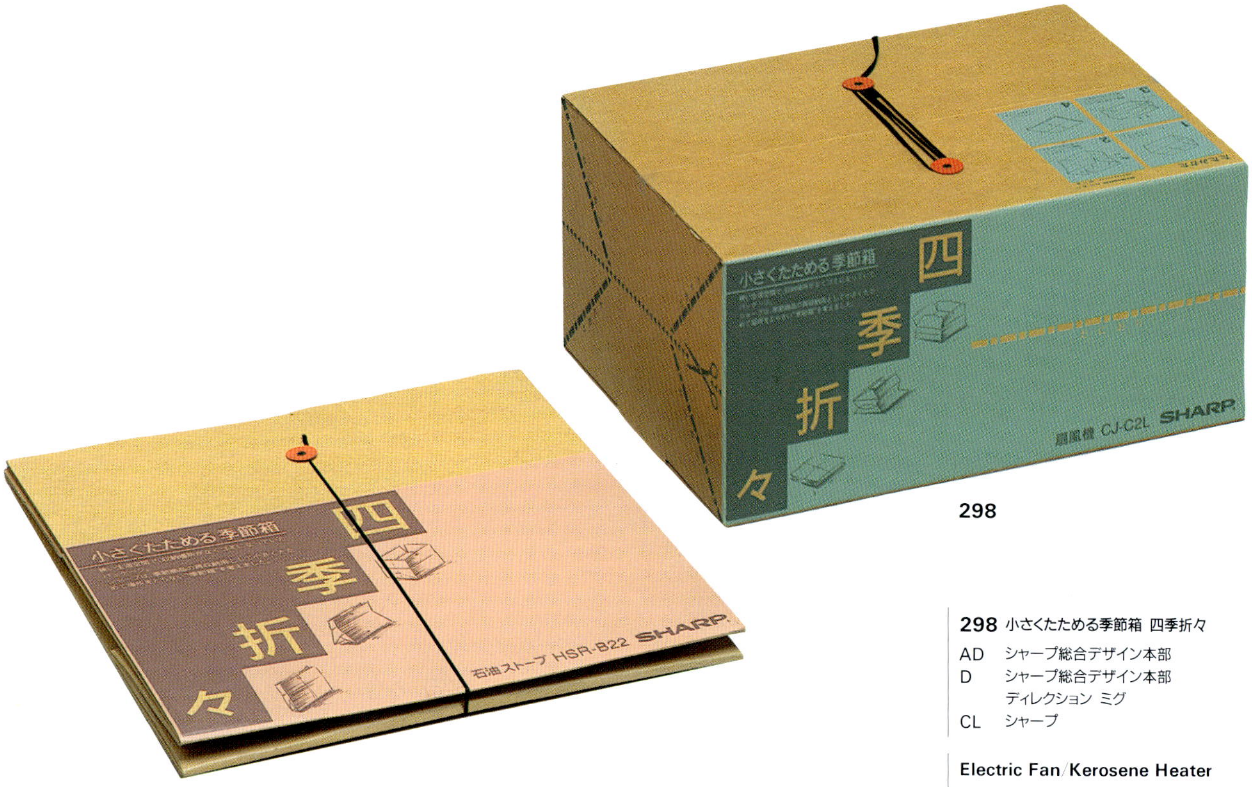

298

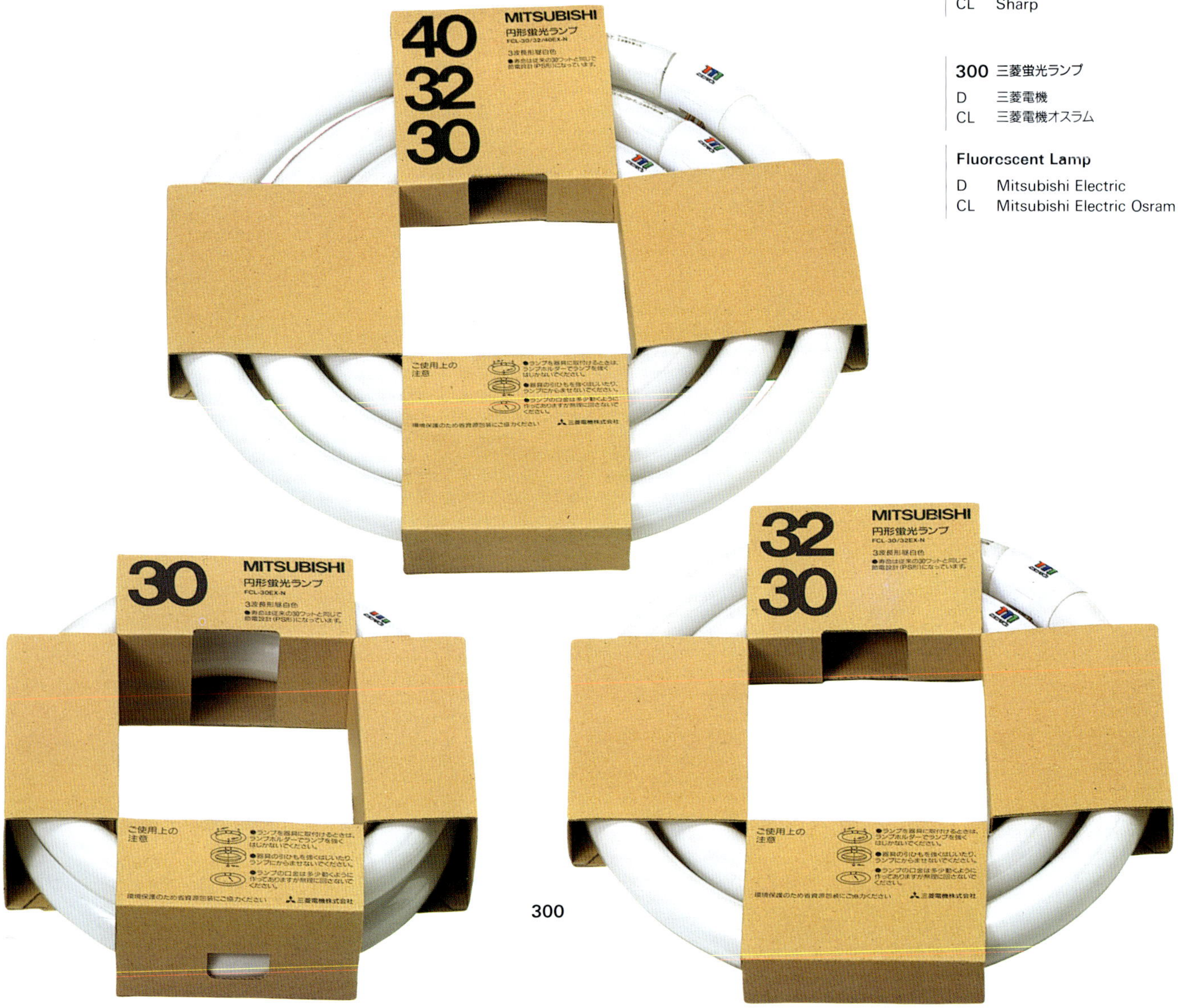

300

298 小さくたためる季節箱 四季折々

AD　シャープ総合デザイン本部
D　シャープ総合デザイン本部
　　ディレクション ミグ
CL　シャープ

Electric Fan/Kerosene Heater

AD　Corporate Design Center, Sharp
D　Corporate Design Center, Sharp
　　Direction Mig
CL　Sharp

300 三菱蛍光ランプ

D　三菱電機
CL　三菱電機オスラム

Fluorescent Lamp

D　Mitsubishi Electric
CL　Mitsubishi Electric Osram

299 パッと開いて、アイロンかけちゃお!

AD　シャープ電化デザインセンター
D　シャープ電化デザインセンター
CL　シャープ

Steam Iron

AD　Home Appliances Design Center, Sharp
D　Home Appliances Design Center, Sharp
CL　Sharp

299

301

302

301 ドルツ

PL 松下電工宣伝部
松村和美
AD TCD
D 藤脇慎吾
野田峰子
CL 松下電工

Health-Care Products

PL Advertising Divisions, Matsushita Electric Works
Kazumi Matsumura
AD TCD
D Shingo Fujiwaki
Mineko Noda
CL Matsushita Electric Works

302 フマキラー ベープリキッドシリーズ

PL　蟹江明伸
　　小川 章
AD　蟹江明伸
D　佐々木佳代
　　細沼宏宣
Copy　竹島 靖
PL　眞保昭男
CL　フマキラー

Electronic Mosquito Repeller

PL　Harunobu Kanie
　　Akira Ogawa
AD　Harunobu Kanie
D　Kayo Sasaki
　　Hironobu Hosonuma
Copy　Yasushi Takeshima
PL　Akio Shimpo
CL　Fumakilla

303 大曲帖

AD　藤井陽一郎
D　前川恵子
　　上野小夜子
Copy　相田たけし
P　藤井陽一郎
CL　CBSソニー

Compact Disc

AD　Yoichiro Fujii
D　Keiko Maekawa
　　Sayoko Ueno
Copy　Takeshi Aida
P　Yoichiro Fujii
CL　CBS Sony

303

304 液晶カラーテレビ **4E-L1**

AD　シャープ電子機器事業本部デザインセンター
D　シャープ電子機器事業本部デザインセンター
CL　シャープ

Liquid Crystal Television

AD　TV and Video Systems Group Design Center, Sharp
D　TV and Video Systems Group Design Center, Sharp
CL　Sharp

304

305

305 バブルジェットプリンタBJ-10e

PL	キヤノン周辺機器デザイン部
AD	犬塚達美
D	犬塚達美
	佐藤勝則
P	新崎紳一郎
CL	キヤノン

Printer

PL	Industrial Design Dept-Peripherals Development Center, Canon
AD	Tatsumi Inuzuka
D	Tatsumi Inuzuka
	Katsunori Sato
P	Shin'ichiro Arasaki
CL	Canon

306

307

306 廃棄物処理を考慮した
パソコン・ワープロパッケージ

CD　平井文夫
AD　真崎太郎
D　松尾 茂
CL　東芝

**Personal Computer/
Word Processor**

CD　Fumio Hirai
AD　Taro Masaki
D　Shigeru Matsuo
CL　Toshiba

307 ワンダーメガ

AD　日本ビクターデザインセンター
D　日本ビクターデザインセンター
グラフィックデザイン室
DL　日本ビクター

Computer

AD　Design Center,
Victor Company of Japan
D　Graphic Design Dept., Design
Center, Victor Company of Japan
CL　Victor Company of Japan

GIFT FOR YOU VOGEL
The delicious cake of gorgeousness and beauty building up more pleasure.
FOR YOU GIFT AMERICAN BEAR
The delicious cake of gorgeousness and beauty building up more pleasure.
PRESENT FOR YOU.PRESENT FOR YOU.PRESENT FOR YOU

308

308 アメリカンベア、フォーゲル

AD　本州製紙パッケージングセンター
D　アートスタイルマーケット
CL　柳井紙工

Gift Box

AD　Honshu Packaging Center
D　Art Style Market
CL　Yanai

309 レ・エッセンシャル

AD　中川憲造
D　延山博保
　　森上 暁
　　前川浩美
CL　信光社

Jewelry

AD　Kenzo Nakagawa
D　Hiroyasu Nobuyama
　　Satoshi Morikami
　　Hiromi Maekawa
CL　Shinkosha

309

310 キャリーバッグ、包装紙

CD　杉野 壮
AD　南部俊安
D　南部俊安
CL　松下電工

Paper Bag/Wrapping Paper

CD　Tsuyoshi Sugino
AD　Toshiyasu Nanbu
D　Toshiyasu Nanbu
CL　Matsushita Electric Works

310

311 ショッピングバッグ
AD 近藤一弥
D 近藤一弥
CL 草月アートプランニング

Paper Bag
AD Kazuya Kondo
D Kazuya Kondo
CL Sogetsu Art Planning

312 富士銀行ペーパーバッグ
AD 佐藤 卓
D 佐藤 卓
CL 富士銀行

Paper Bag
AD Taku Sato
D Taku Sato
CL Fuji Bank

313 キャリーバッグ
CD 杉野 壮
AD 南部俊安
D 南部俊安
CL 松下電工

Paper Bag
CD Tsuyoshi Sugino
AD Toshiyasu Nanbu
D Toshiyasu Nanbu
CL Matsushita Electric Works

311

312

313

314 ラフォーレ ショッピングバッグ
AD 大貫卓也
D 島林裕一
P 白鳥真太郎
CL ラフォーレ原宿

Paper Bag
AD Takuya Onuki
D Yuichi Shimabayashi
P Shintaro Shiratori
CL Laforet Harajuku

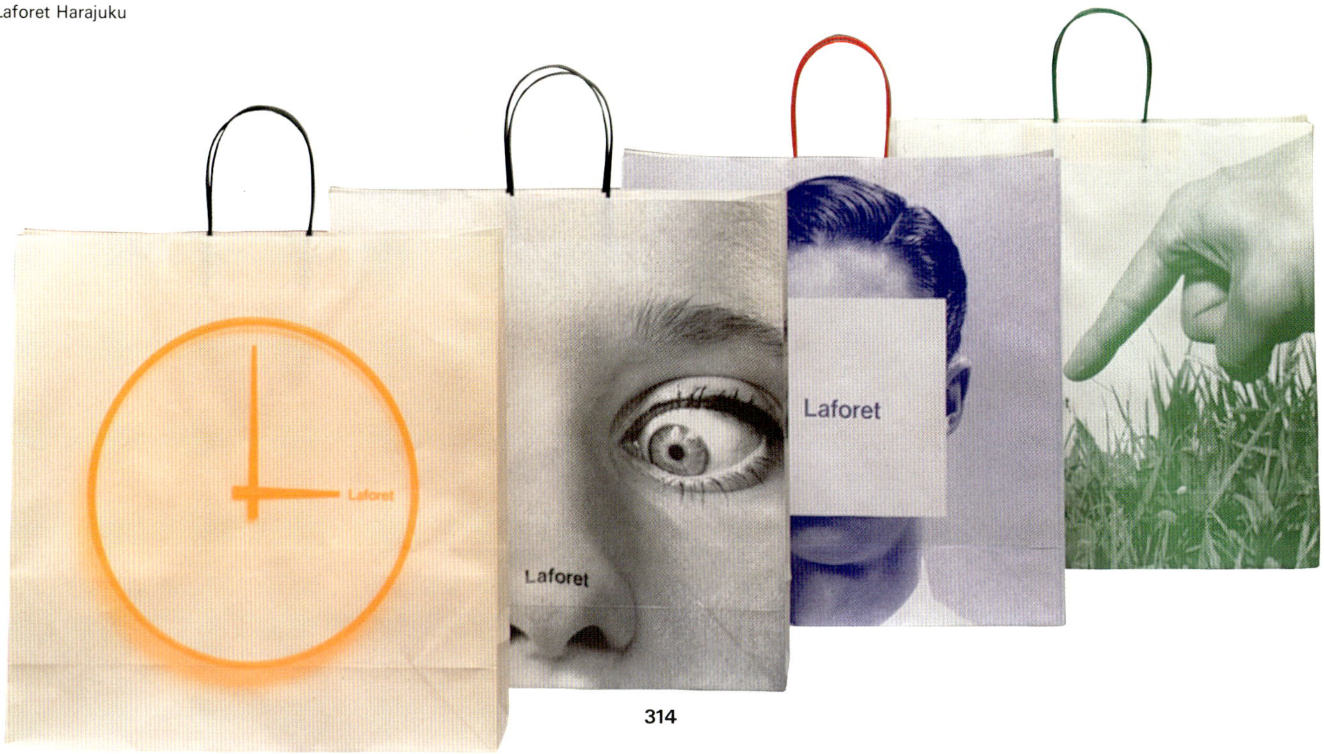

314

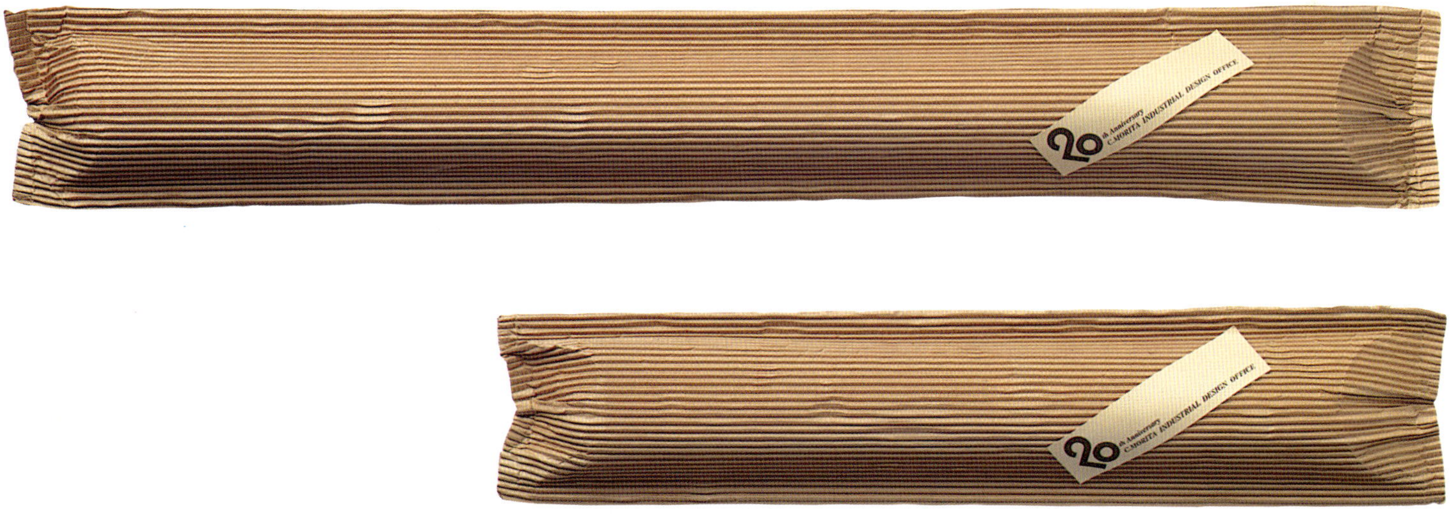

315

315 アジャストケースパッケージ
AD 奥村昭夫
D 南 勝治
CL 森田力工業デザイン研究所

Adjustable Case
AD Akio Okumura
D Katsuji Minami
CL C. Morita Industrial Design Office

316 ヨックモック ショッピングバッグ

AD 近藤 司
松原章二
D 川路ヨウセイ
小澤満佐子
CL ヨックモック

Paper Bag

AD Tsukasa Kondo
Shoji Matsubara
D Yosei Kawaji
Masako Ozawa
CL Yoku Moku

317 キャリーバッグ

CD 杉野 壮
AD 南部俊安
田中雅資
D 南部俊安
CL 松下電工労働組合

Paper Bag

CD Tsuyoshi Sugino
AD Toshiyasu Nanbu
Masashi Tanaka
D Toshiyasu Nanbu
CL Matsushita Electric Works
Workers' Union

316

317

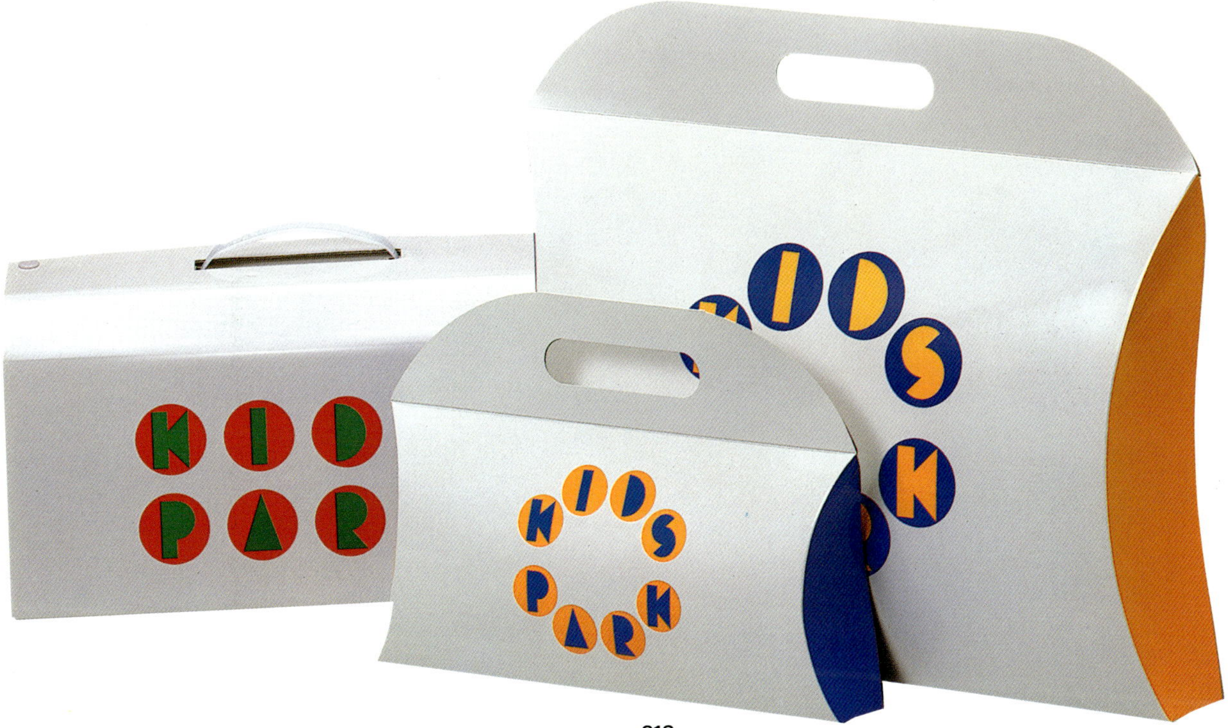

318

319

318 Kids Park

AD 廣村正彰
D 小島利之
Const 本田和男
CL 西友

Clothes

AD Masaaki Hiromura
D Toshiyuki Kojima
Const Kazuo Honda
CL The Seiyu

319 キッコーマン ワインバッグ

AD 那波多目康子
D 那波多目康子
Const 那波多目康子
IL 村上みどり
CL キッコーマン

Wine Carry Bag

AD Koko Nabatame
D Koko Nabatame
Const Koko Nabatame
IL Midori Murakami
CL Kikkoman

320 シャンドーレ バレンタイン
'92包装紙

AD　宇田 優
D　奥田一明
CL　シャンドーレ

Chocolate

AD　Masaru Uda
D　Kazuaki Okuda
CL　Champsdorés

320

321

321 駿河屋
リサイクルペーパー **菓紙**

CD　前田和紀
CL　駿河屋

Japanese Sweets

CD　Kazuki Maeda
CL　Souhonke Surugaya

322

322 ロシナンテ包装紙

AD　上本一寿
D　上本一寿
LO　上本一寿
CL　ロシナンテ

Confectionery

AD　Kazutoshi Uemoto
D　Kazutoshi Uemoto
LO　Kazutoshi Uemoto
CL　Rossinante

323

323 男爵コロッケ

PL　長谷川和正
AD　廣畑勝雄
D　廣畑勝雄
IL　スージィ甘金
Copy　河村 悟
CL　ローソン

Brown Bag

PL　Kazumasa Hasegawa
AD　Katsuo Hirohata
D　Katsuo Hirohata
IL　Suzy Amakane
Copy　Satoru Kawamura
CL　Lawson

324

324 乳製品ギフトシリーズ

AD シヴァデザイン事務所
D 清水直明
CD TPパック
CL 北海道乳業

Dairy Products

AD Siva Design Office
D Naoaki Shimizu
CD TP Pack
CL Hokkaido Nyugyo

325 エピスリ 花抄院

AD 荒木茂野
D 荒木茂野
IL 荒木茂野
CL 長﨑堂

Cake

AD Shigeno Araki
D Shigeno Araki
IL Shigeno Araki
CL Nagasakido

331 ロマーナパスタセット
D　本州製紙パッケージングセンター
CL　昭和産業

Assorted Pasta
D　Honshu Packaging Center
CL　Showa Sangyo

332

333

332 ヘルシーフード ギフトパッケージ
AD　朝妻けい子
D　朝妻けい子
CL　岩瀬産業

Snack
AD　Keiko Asazuma
D　Keiko Asazuma
CL　Iwase Sangyo

333 日清製粉ママーギフト
AD　竹内 豊
D　川路ヨウセイ
　　宮本泰志
CL　日清製粉

Pasta/Sauce
AD　Yutaka Takeuchi
D　Yosei Kawaji
　　Yasushi Miyamoto
CL　Nisshin Flour Milling

334

334 サボンドール
CAD 松原正毅
D ヘルムート・シュミット
CL 資生堂

Soap
CAD Masaki Matsubara
D Helmut Schmid
CL Shiseido

337

337 タッパーウェア フリーザーメイト
AD 村 武史
D 鈴木良彦
P 鈴木写真事務所
CL 日本タッパーウェア

Plastic Containers
AD Takeshi Mura
D Yoshihiko Suzuki
P Suzuki Photo
CL Tupperware Japan

335 天然調味料セット

AD GKグラフィックス
D GKグラフィックス
CL 西武百貨店

Seasoning

AD GK Graphics
D GK Graphics
CL The Seibu Department Stores

336 峰鶴之揃／豊醇之揃／萠味之揃／迎春之揃

AD 前田和紀
D 前田和紀
Copy 西田太一
CA 前田誠一
CL 駿河屋

Japanese Sweets

AD Kazuki Maeda
D Kazuki Maeda
Copy Taichi Nishida
CA Seiichi Maeda
CL Souhonke Surugaya

335

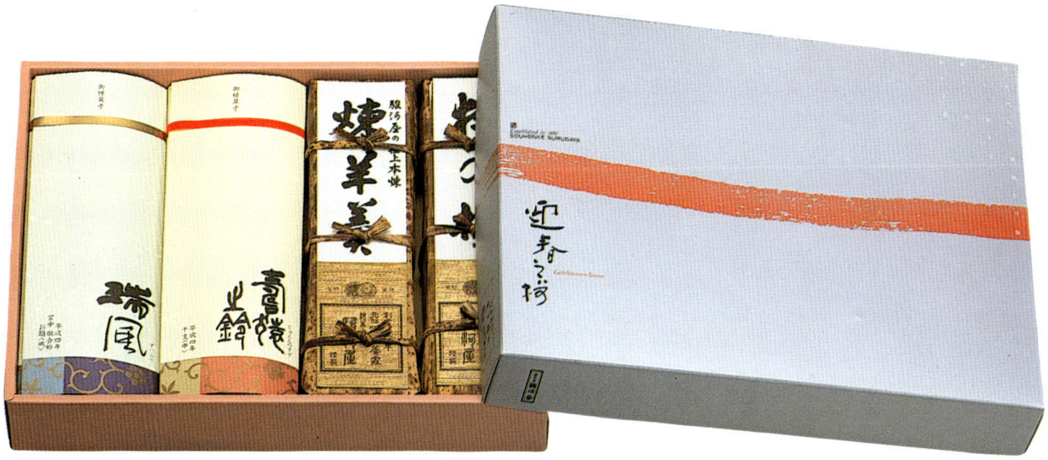

336

338

338 マイディナー
CD 江口 收
AD 江口 收
D 立沢トオル
P 梶 洋哉
Copy 山下みどり
PR 山下みどり
CL 明治屋

Stew
CD Osamu Eguchi
AD Osamu Eguchi
D Toru Tachizawa
P Hiroya Kaji
Copy Midori Yamashita
PR Midori Yamashita
CL Meidi-Ya

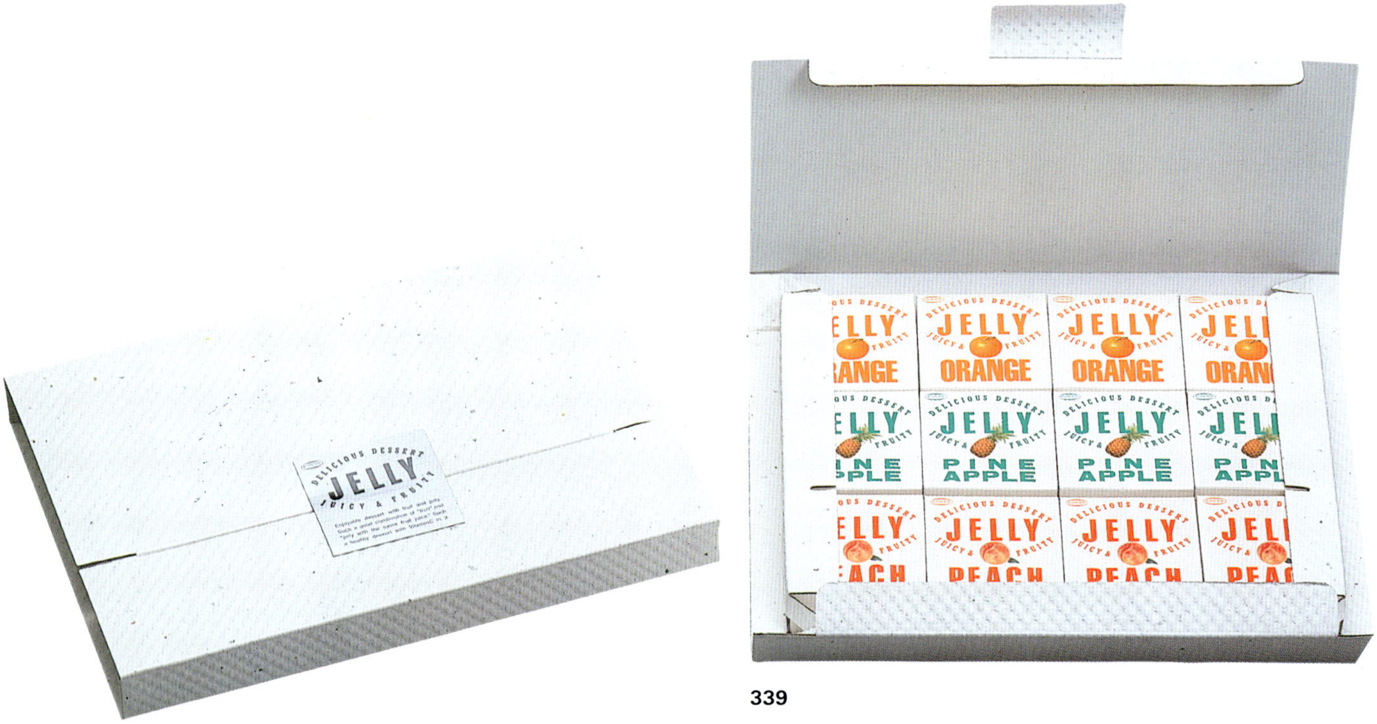

339

339 フルーツゼリーギフトセット
PL ラウンド
CD 吉相博正
AD 和田龍弥
D 和田龍弥
CL 日清製粉

Dessert Gift
PL Round
CD Hiromasa Kisso
AD Tatsuya Wada
D Tatsuya Wada
CL Nisshin Flour Milling

340 **Ajinomoto Gift**
CD 梅田悦史
CL 味の素

Oil/Used Oil Hardener
CD Etsufumi Umeda
CL Ajinomoto

340

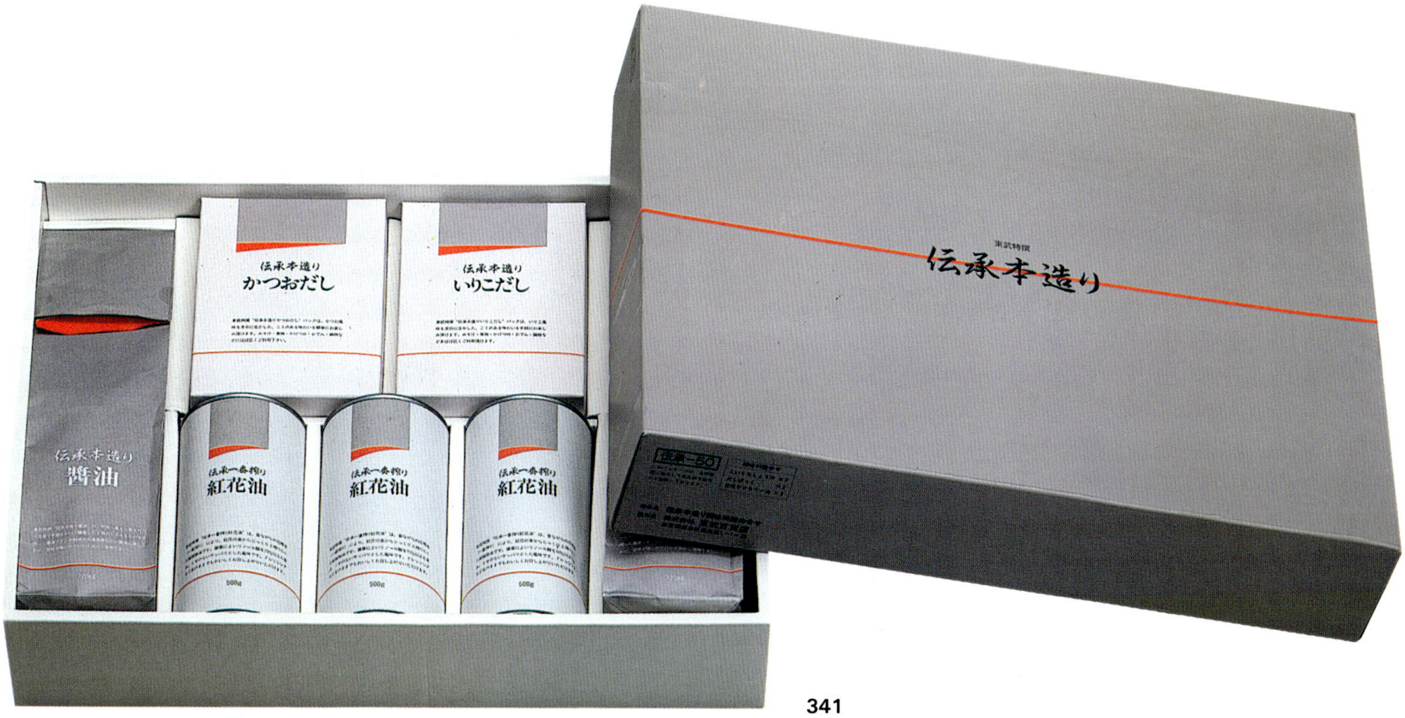

341

341 伝承本造り 調味料セット
AD GK京都
D GK京都
CL 東武百貨店

Seasoning
AD GK Kyoto
D GK Kyoto
CL Tobu Department Store

342

342 駿河屋カートン

AD 前田和紀
D 山地良子
CL 駿河屋

Japanese Sweets

AD Kazuki Maeda
D Yoshiko Yamaji
CL Souhonke Surugaya

343 東芝クリーナー

CD 平井文夫
AD 居阪みき代
D 高原文彦
CL 東芝

Cleaner

CD Fumio Hirai
AD Mikiyo Isaka
D Fumihiko Takahara
CL Toshiba

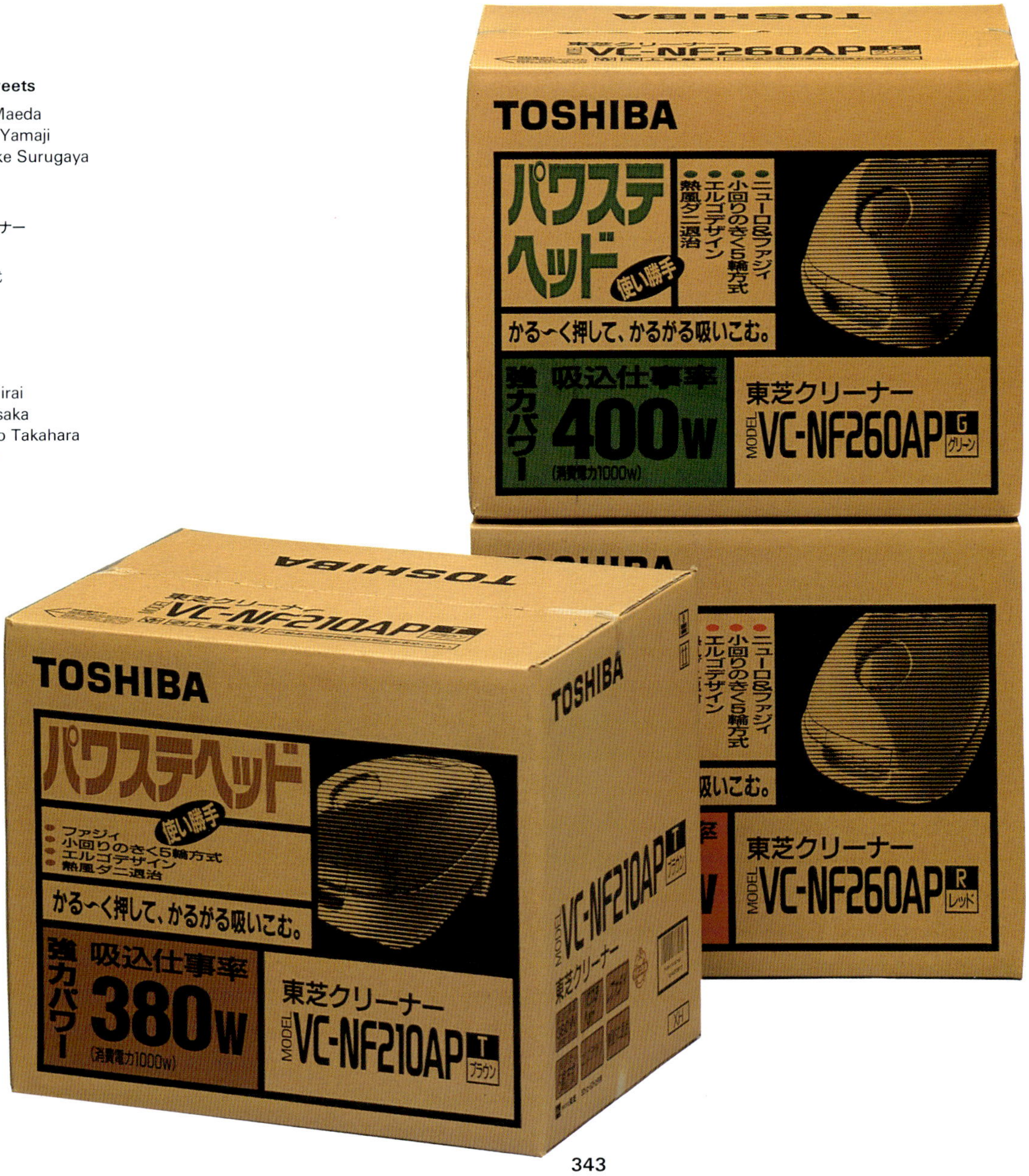

343

344

344 東芝カラーテレビ、東芝ビデオカセットレコーダー

CD 平井文夫
AD 真崎太郎
D 松尾 茂
小川洋子
CL 東芝

Color TV/Video Cassette Recorder

CD Fumio Hirai
AD Taro Masaki
D Shigeru Matsuo
Yoko Ogawa
CL Toshiba

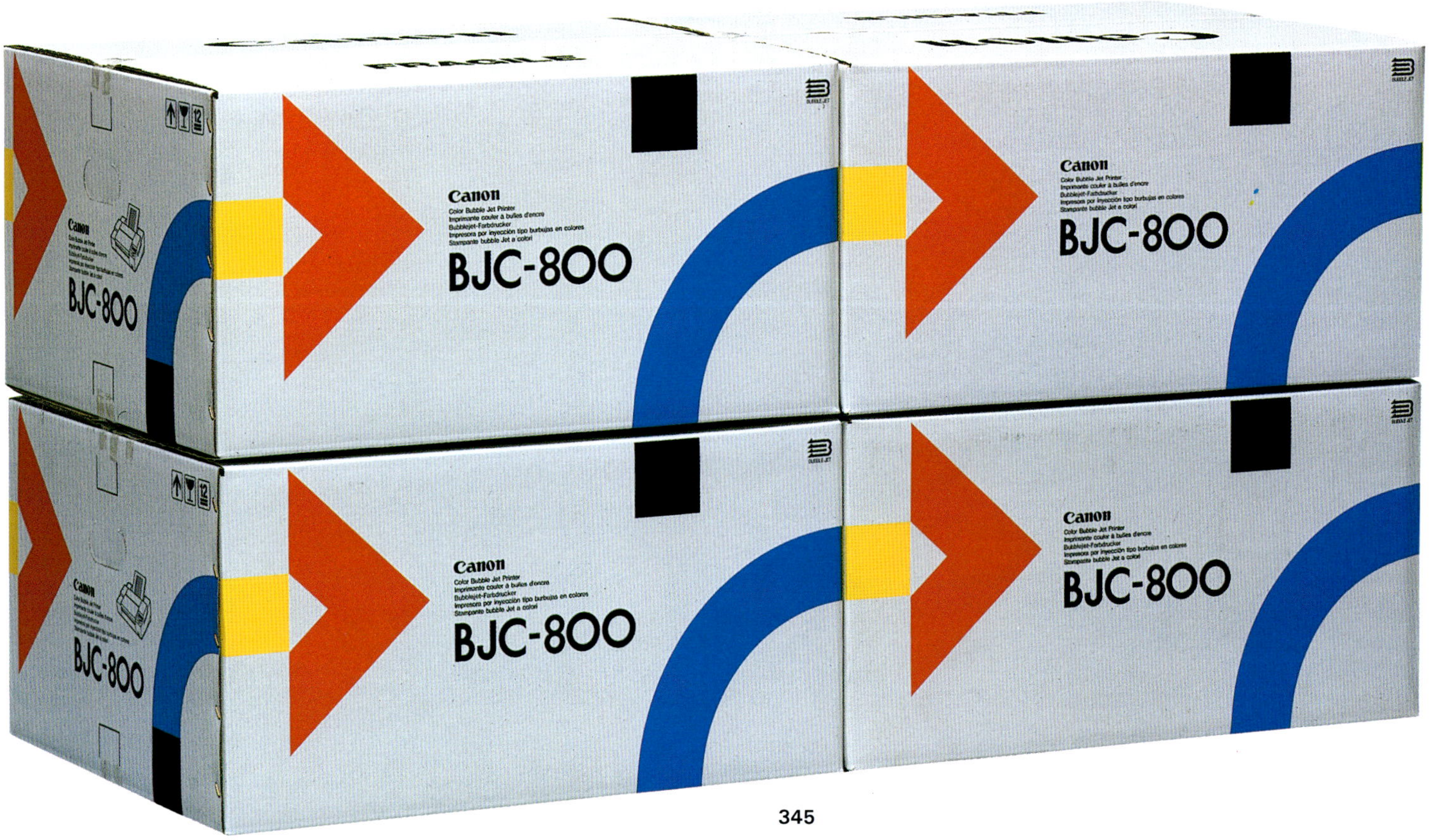

345

345 カラーバブルジェットプリンタ**BJC-800**

PL　キヤノン周辺機器デザイン部
AD　犬塚達美
D　犬塚達美
　　佐藤勝則
CL　キヤノン

Printer

PL　Industrial Design Dept., Peripherals Development Center Canon
AD　Tatsumi Inuzuka
D　Tatsumi Inuzuka
　　Katsunori Sato
CL　Canon

346 高知県野菜、果物段ボール

AD　本州製紙パッケージングセンター
CL　高知県園芸連

Vegetable

AD　Honshu Packaging Center
CL　Kochi Prefecture Federation of Horticultural Agricultural Cooperatives

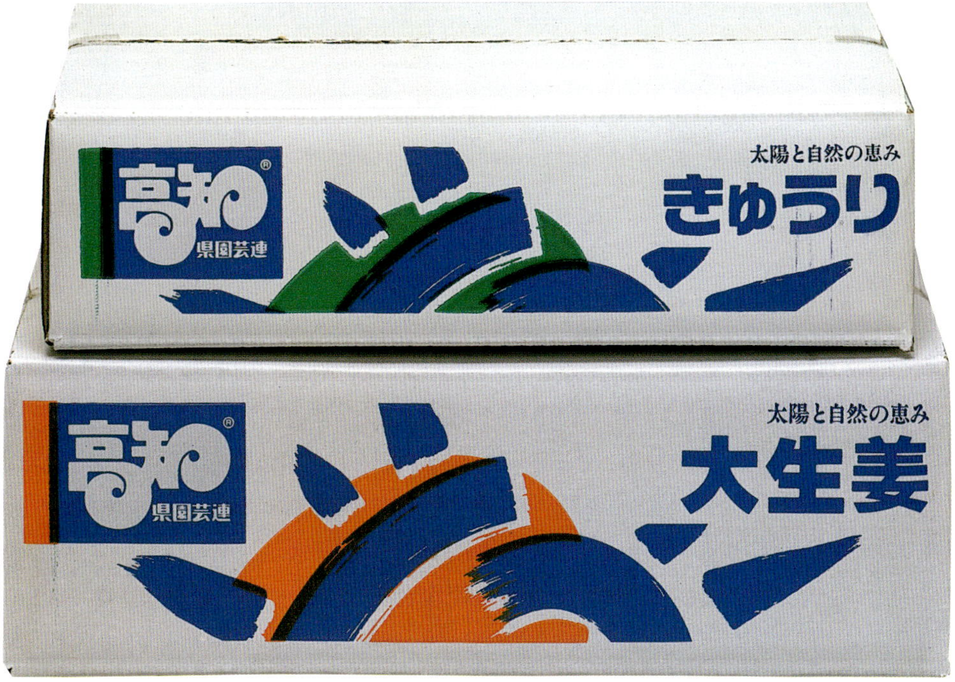

346

347

347 長野県野菜段ボール

AD　本州製紙パッケージングセンター
CL　長野県経済連

Vegetable

AD　Honshu Packaging Center
CL　Nagano Prefecture Federation of Agricultural Cooperatives

348

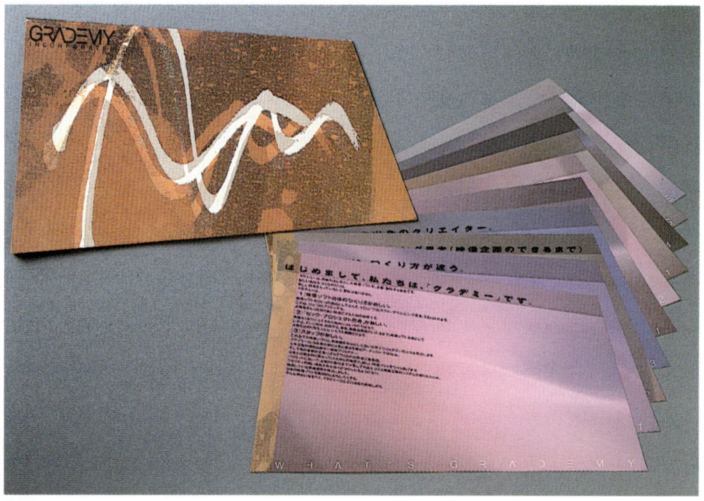

349

350

348 コミーツーリスト封筒
AD　竹智 淳
D　竹智 淳
CL　コミーツーリスト

Envelope
AD　Jun Takechi
D　Jun Takechi
CL　Kommy Tourist

349 グラデミー会社案内
AD　竹智 淳
D　竹智 淳
CL　グラデミー

Company Brochure
AD　Jun Takechi
D　Jun Takechi
CL　Grademy

350 グラデミー封筒
AD　竹智 淳
D　竹智 淳
CL　グラデミー

Envelope
AD　Jun Takechi
D　Jun Takechi
CL　Grademy

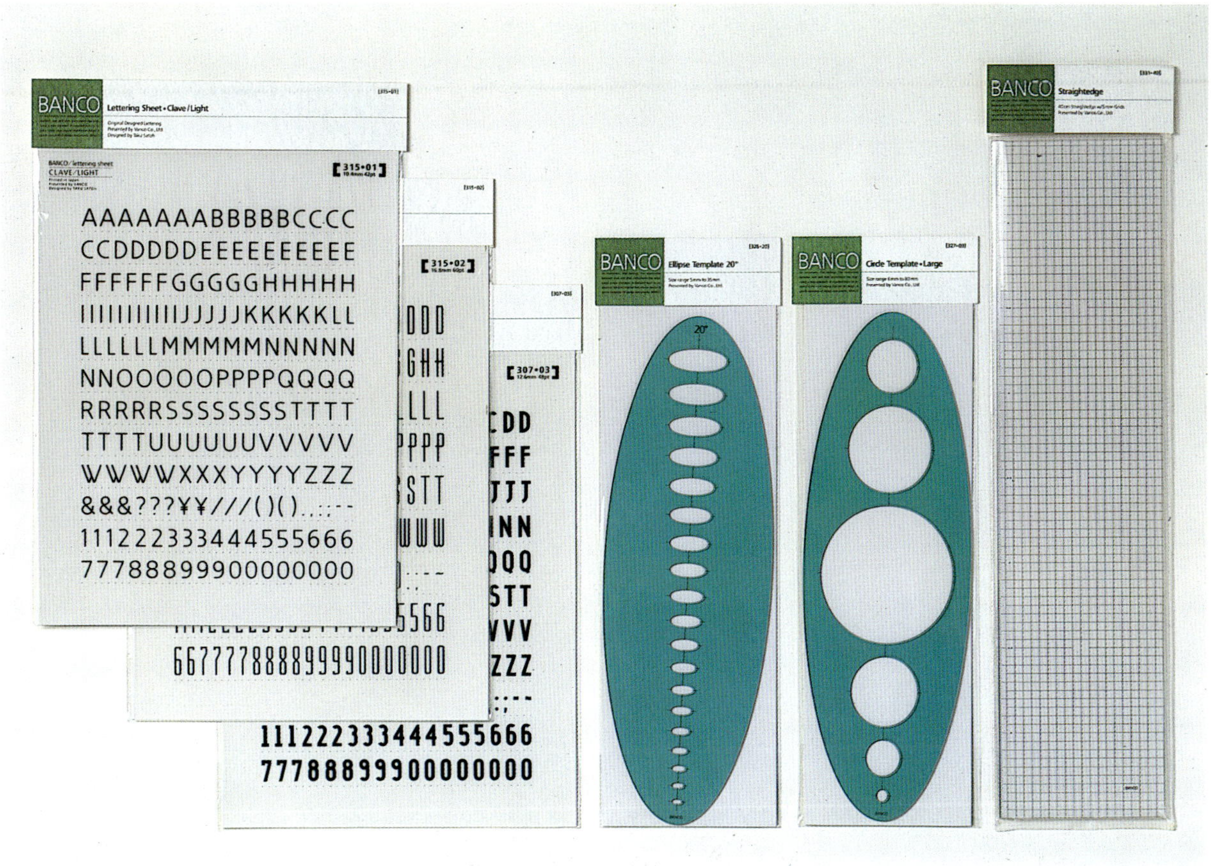

351

351 バンコデザインツール
AD 佐藤 卓
D 佐藤 卓
CL バンコ

Design Tool
AD Taku Sato
D Taku Sato
CL Vanco

352 視覚・解体新書
AD 松井桂三
D 松井桂三
荒木優子
CL 岡本印刷工業

Promotional Tool
AD Keizo Matsui
D Keizo Matsui
Yuko Araki
CL Okamura Printing Industry

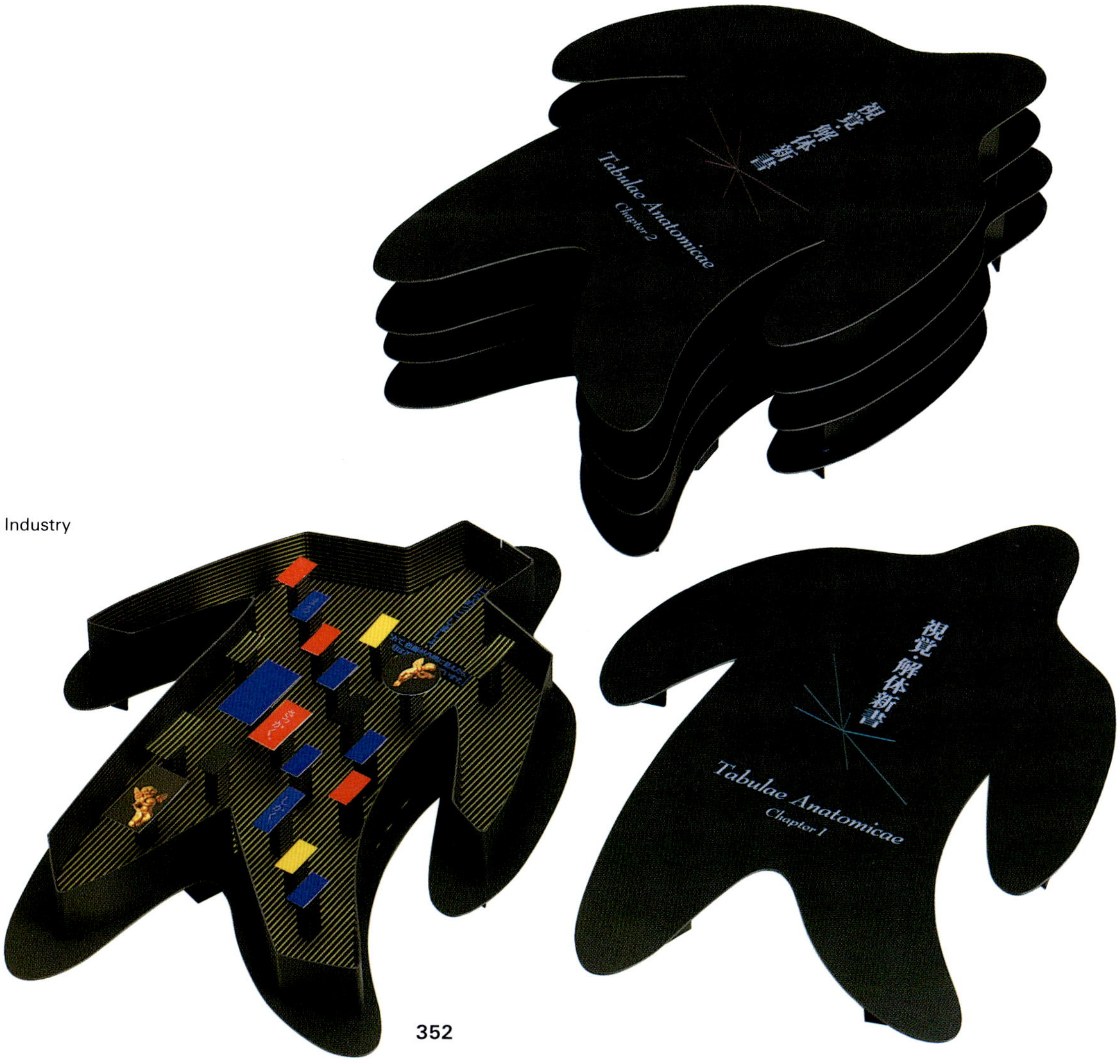

352

353 コスモオイル リオ
AD 和田 亨
D 松尾高明
CL コスモオイル

Oil
AD Toru Wada
D Koumei Matsuo
CL Cosmo Oil

353

354 環境優選シリーズ
AD 太田徹也
Copy 川上嘉端
AG 大日本印刷PAC
CL 西友

Daily Goods Variety
AD Tetsuya Ota
Copy Kazui Kawakami
AG Dai Nippon Printing PAC
CL The Seiyu

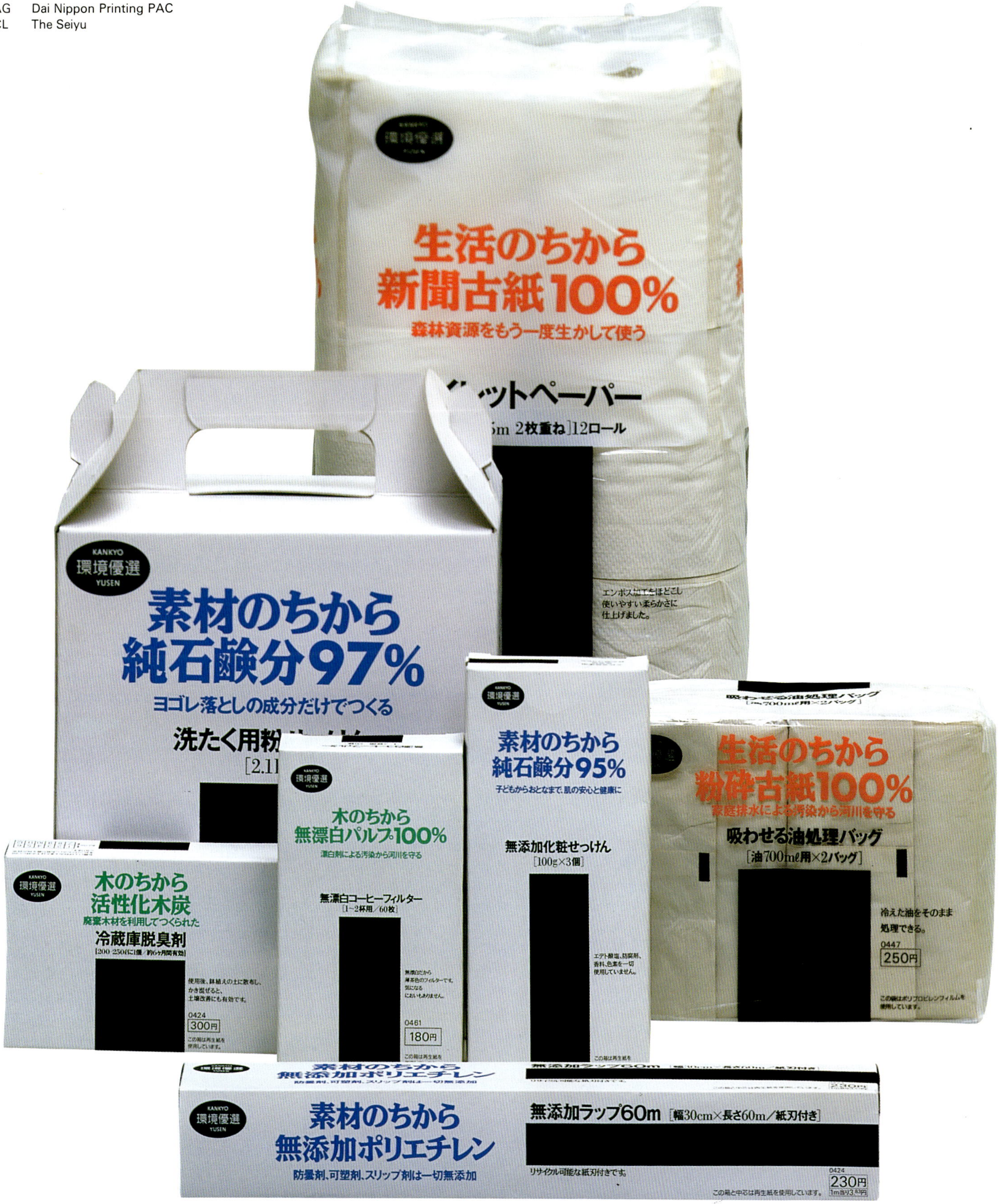

354

特集：

その時代と心に刻み込まれたパッケージ

Special Feature:
Impressive Packages Reflecting Current Needs

時代は今、地球環境問題に直面し、パッケージのあり方も大きな転換期を迎えている。思い起こしてみると、昭和30年代、バラ売りの菓子や量り売りの味噌・醤油などがまたたくまに姿を消した。以来、大量生産・大量消費の時代に入り、商品がパッケージ化された。そして、現在の我々の周囲を眺めてみれば、様々なパッケージデザインが日常の生活環境をかたちづくる大きな要素となっているのがよくわかる。

日頃、売り手の側から語られ、デザインの優劣で語られることの多いパッケージデザインを、我々の記憶をたよりに拾い直してみることで、時代の流れや生活の中にパッケージがどのように根を下ろし、生き続けてきたのかを振り返ってみたい。その中から、ゴミや資源としてのパッケージだけではなく、我々の未来の生活を表現し、演出してゆくデザインとしての側面も見えてくるのではないだろうか。

この特集を組むにあたり、「戦後の日本のパッケージデザインでインパクトのあったパッケージデザイン、記憶にすりこまれているパッケージデザインは何か」というアンケートを、日本パッケージデザイン協会に所属するパッケージデザイナーを対象に行った。その結果、輸入品も含めて、数百のパッケージデザインが寄せられた。ここに掲載したのは、その一部である。

特集編集委員

青木茂吉 1933年生まれ
株式会社デザインネットワーク取締役社長

金子修也 1937年生まれ
株式会社GKグラフィックス取締役社長

大森重志 1939年生まれ
サントリー株式会社デザイン部長

高橋　敏 1948年生まれ
株式会社イフカンパニー代表取締役

犬塚達美 1950年生まれ
株式会社犬塚デザイン事務所代表取締役

Committee Members

Shigeyoshi Aoki born in 1933
President of Design Network Co., Ltd.

Shuya Kaneko born in 1937
President of GK Graphics Inc.

Shigeshi Omori born in 1939
General Manager of Design Department, Suntory Ltd.

Ving Takahashi born in 1948
President of IFF Company Inc.

Tatsumi Inuzuka born in 1950
President of Inuzuka Design Office

凡例 **Legent**

[1] 商品名 Title of Product
[2] 発売年 Year of Sale(in Japan)
[3] デザイン年(掲載図版) Year of Design(of Photo)
[4] 製造・販売 Manufacturer and Distributor

[1] カルピス Calpis [2] 1919(大正8) [3] 1949(昭和24) [4] カルピス食品工業 The Calpis Food Industry

［1］サントリーウイスキー Suntory Whisky ［2］1937（昭和12） ［3］1937（昭和12） ［4］サントリーSuntory

[1] キリンラガービール Kirin Lager Beer [2] 1889(明治22) [3] 1890(明治23) [4] キリンビール Kirin Brewery

[1] カップヌードル Cup Noodle　[2] 1971(昭和46)　[3] 1971(昭和46)　[4] 日清食品 Nissin Food Products

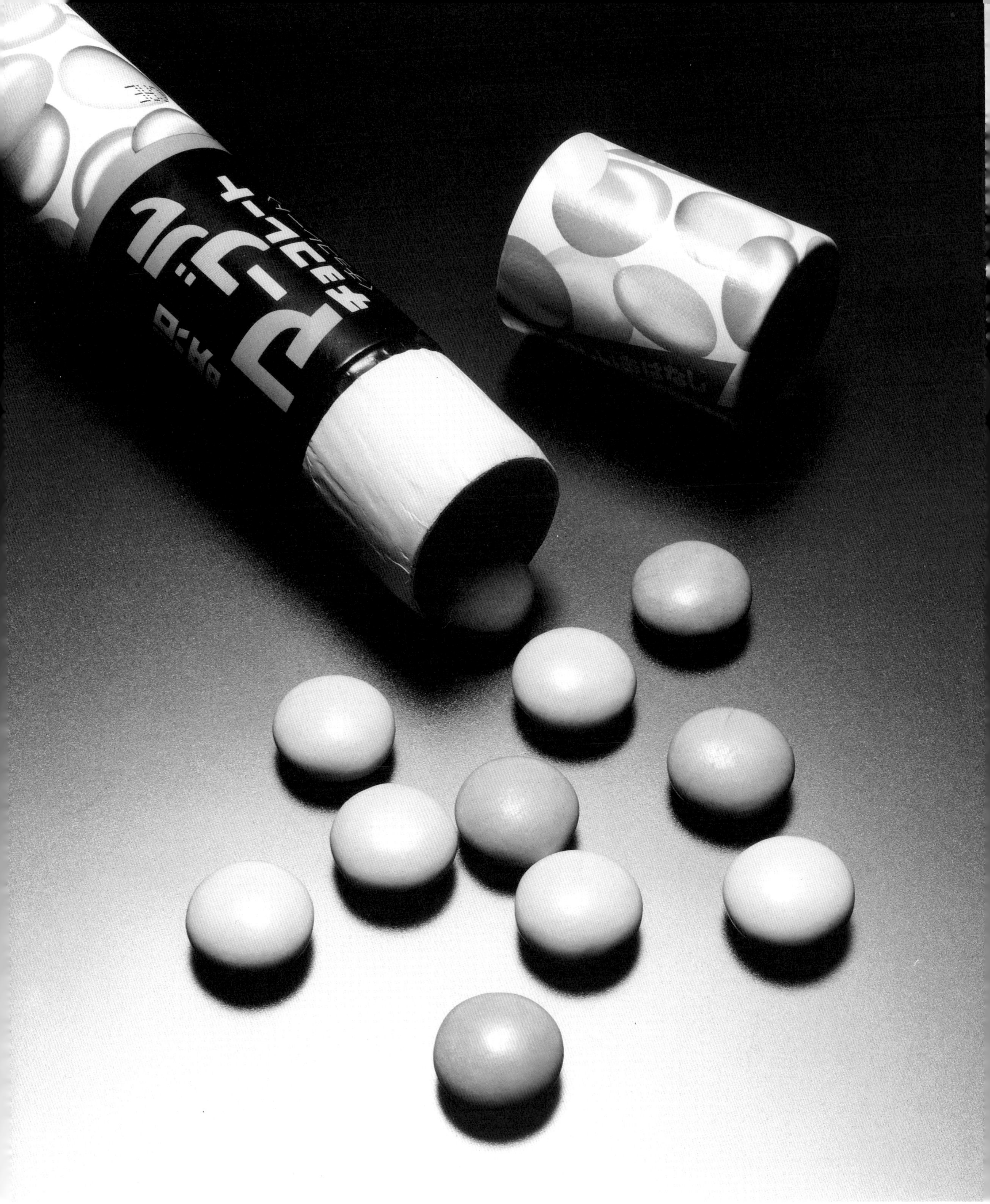

[**1**] 明治マーブルチョコレート Meiji Marble Chocolate [**2**] 1961（昭和36） [**3**] 1961（昭和36） [**4**] 明治製菓 Meiji Seika Kaisha

[1] MG5 ヘアトニック MG5 Hair Tonic [2] 1967(昭和42) [3] 1967(昭和42) [4] 資生堂 Shiseido

〔**1**〕マジックインキ Magic Ink 〔**2**〕1952(昭和27) 〔**3**〕1952(昭和27) 〔**4**〕寺西化学工業 Teranishi Chemical Industry

［1］三越の包装紙 Mitsukoshi Wrapping Paper ［2］1951（昭和26） ［3］1948（昭和23） ［4］三越 Mitsukoshi

[1] 化粧石鹼カウブランド赤箱 Cow Brand Beauty Soap(Red Box)　[2] 1949(昭和24)　[3] 1986(昭和61)　[4] 牛乳石鹼共進社 Cow Brand Soap Kyoshinsha

[1] 金鳥の渦巻 Kincho's Mosquito Coil [2] 1895(明治28) [3] 1955(昭和30) [4] 大日本除虫菊 Ueyamayei's Japan Insecticide MFG

●甘味への憧れと水玉模様——カルピス

青木　パッケージデザインに対する考え方がしっかりしてきたのは、昭和30年代ですが、そのずっと前、大正8年からカルピスがあるんですね。ただのグラフィックデザインとか商品デザインじゃなくて、クシャクシャの柔らかい薄い紙でくるむ、水玉模様の紙でくるむというふうに素材のすがすがしさを出していますね。

大森　当時のお酒もこういう包装だったようですね。

青木　化粧品などでも薄紙でくるむというのがいっぱいありました。それを帯で留めるなどね。大量生産になって機械化できないものは、昭和40年代からどんどんなくなりましたけどね。

犬塚　生き残ったのでしょうね、カルピスは。

大森　カルピスの水玉は強烈で、そのあと森永乳業もこれに続いて、乳製品にはこの記号を使いましたね。

高橋　水玉は、ストライプやチェックと同じく自然現象を意味していますね。カルピスはロゴがなくても水玉だけでいける唯一のデザインですね。

金子　甘味は江戸時代から大衆の憧れ。明治時代も尾を引いて、しかも南方、海外から入ってくるものという憧れがあった。水玉の向うに南洋の光があった。時代の記号ですね。しかも今の時代に残っている。

●インテリアとしての角瓶——サントリーウイスキー

大森　なぜオールドでなく角瓶が選ばれたのかな。

金子　角瓶が発売された昭和12年は僕の生まれ年。角瓶は、昭和前期の中産階級の住宅の応接間や書斎に置かれるものというステイタスを感じますね。オールドになると住宅の中に置くというよりむしろ外で飲むもの、サロンのものですね。

大森　確かにどこの家に行っても角瓶が1本ありましたね。

青木　ショットグラスにご主人がついでくれてね。

金子　玄関、上がり框、中廊下、すぐ左側に応接間。畳の上にアラベスク調の絨毯、そして回転椅子と大きな木製書棚。その中にこの角瓶がある。まさにインテリアグッズですね。

青木　それに対してオールドは、ボトルキープのイメージですね。

金子　トリスバーがサントリーバーになって、棚に並ぶはオールド。そして若かったママがオールド婆さんになっていた(笑)。

犬塚　「角(かく)」というのは商標ではないんですよね。

大森　そうです。周りからいいはじめた。内部で戦略的に「角瓶」としたわけではないんです。

大森　最初の角瓶は薄い2色刷くらいの唐草模様の紙で包んでいて、高級感があった。これを見直そうとして薄い包装紙を復活させてもらったことがありました。紙を通して中身が透けて見えるのがおいしさをものすごく感じさせますね。

犬塚　僕は、亀甲模様が印象に残る。キラキラしていてきれいですね。

青木　瓶の肩が皆、丸くなった時代があるんですよ。大量生産の時代とオイルショックの時代。まず、昭和35年から40年にかけて製瓶にアイエスマシンが導入されるようになったときの肩直し。それからオイルショック。肩を丸くするとガラスの目付けが少なくてすむし、ガラス自体は高いものではないけれど流通・製造の総エネルギーが減りますから。

●インスタント食品革命——カップヌードル

金子　ラーメン、コーヒーは自分の家でつくれないもの、つくりにくいもの

[1] グリコ Glico
[2] 1921(大正10)
[3] 1992(平成4)
[4] 江崎グリコ Ezaki Glico

[1] 森永ミルクキャラメル(復刻版) Morinaga Milk Caramels Classic Version
[2] 1899(明治32)
[3] 1979(昭和57)
[4] 森永製菓 Morinaga

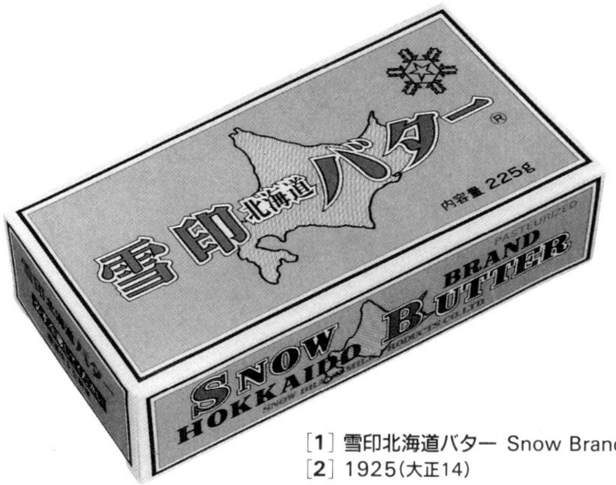

[1] 雪印北海道バター Snow Brand Hokkaido Butter
[2] 1925(大正14)
[3] 1962(昭和37)
[4] 雪印乳業 Snow Brand Milk Products

[1] キユーピーマヨネーズ Q.P.Mayonnaise
[2] 1925(大正14)
[3] 1958(昭和23)
[4] キユーピー Q.P.

だったのが、昭和30年代にインスタントが出てきてショックだった。それがカップヌードルになって、お湯さえあればどこでも食べられることになり、さらに大飛躍しましたね。

大森　はじめのチキンラーメンは強烈だった。中学、高校受験の思い出ですね。ちょっと煮込んだほうがおいしかったね。

金子　カップヌードルの登場で山登りの装備が変わりました。また、ベトナム戦争のときにだいぶ使われたようですね。

大森　戦争の度にパッケージが大きく変わりましたね。まず第1次世界大戦のとき、ほとんど今のパッケージの原型ができたといわれています。保存、輸送、コンパクト化など戦争はパッケージの革命をしたんですね。

金子　日本では米の存在が大きくて、米は、水と火があればなんとか食べられます。でも、パンは戦場ではつくれませんね。それがパッケージ化を促した典型的な要因ですね。それから、缶詰はナポレオンのフランス陸軍で編み出したものと聞いています。

犬塚　カップヌードルの文字を見ると、あの頃「タイポニー」などによって新しいタイプフェイスがどんどん入ってきたということがわかりますね。

金子　いつの間にか「ラーメン」から「ヌードル」になって、外国籍も取った感じ。外国でコマーシャルもありましたね。

高橋　盛んなCM戦略で、CMとパッケージの結びつきの力をとても感じました。

●変わらない価値観に対する安心感———キリンラガービール

大森　キリンラガーは、ながくデザインが変わっていない。ラベルにアルミホイルを使っていて、高級感があって、伝統的で、今ではできないデザインですよね。

金子　変わらない価値観に対する安心感がある。サントリーの角や金鳥蚊取り線香も同じだね。

犬塚　最近缶ビールがポピュラーになってきていますね。

大森　オケイジョンによるのでしょう。生活空間のこともありますが……。瓶というのは高級感と存在感がありますね。

金子　それになんといっても麒麟のキャラクターがすごいですね。麒麟は中央アジアで生まれたシンボルですね。大陸的スケールがある。

大森　やはり変えずにきたのがよいですね。一時は、他に新しいものが出てきて、キリンはダサイというイメージがあったはずですが、アサヒがレイモンド・ローウィ、サッポロがウォルター・ランドを起用したとき新鮮な感じがしたけれど、長続きしなかったですね。

金子　キリンのラベルは紋章的なので強いですね。最近ヨーロッパなどで紋章の持つ力を見直す傾向が出てきていますね。

青木　でも、ヨーロッパでは紋章が沢山使われすぎる気もします。ウイスキー、煙草、男性用化粧品など。うまい整理が必要かもれませんね。

●高度成長期のお菓子———明治マーブルチョコレート

犬塚　マーブルチョコレートの原型はアメリカのm&m'sでしょうけれど、発売当時は、その派手さにわくわくして、とても嬉しかったですね。

金子　駄菓子的要素のパッケージ化でしょうか。

青木　今のものはチョコレートの色が天然色素になって色が地味になりましたね。

大森　上原ゆかりのテレビの宣伝もかわいかったね。

青木　明治製菓は、チョコレートを強力に宣伝する一方で、マーブルチョコレートを単品で大きく宣伝しましたね。高度成長の時期なのでそれができたのでしょう。

犬塚　団塊の世代の食べ盛りの頃でしたね。当時30円。今は100円。シールのおまけも良かった。

大森　あの頃は真面目にデザインしていた感じがします。お菓子でも大人向きのデザインだった。今、お菓子はキャラクターだけの世界になってしまっている傾向もみられます。

高橋　キャラメルは、煙草の代用だという歌い文句でしたよね。それから、森永の黄色、グリコの赤というのが印象的ですね。

青木　それともっと昔のお菓子というと、駄菓子が多くて、こういうブランドものは高かったから、遠足の思い出に重なりますね。

●自分たちの世代の男性化粧品———MG5ヘアトニック

青木　昭和38年に細口のガラス瓶で2種のMG5が出ていたのですが、4年後にこのデザインを22品種全商品そろえて出しました。

大森　発売のときキャップの大きさにびっくりしました。

青木　この頃PP（ポリプロピレン）が値段も安くなって、大量生産ができて品質がよくなったのです。成形性がよく、金型もよいので誤差がなかった。それでパッキングを取ることができて、その分がキャップに回りました。キャップの規格と瓶の規格を合わせ、キャップを大きくすることで新鮮さを出しながら、ボトルや瓶形は全部丸瓶で、大量生産向きにやりました。

大森　この画期的な大きさのキャップが与えた影響は大きいですね。うちでもカスタムというウイスキーをつくったとき、グラスつきで瓶とキャップを一体化して、帽子のようにかぶせました。

青木　MG5は、素材革命の中でつくられたといえますね。シュリンク、銀の印刷、キャップの拡大など。またテレビや新聞の宣伝を最初から真面目に考えていましたね。

大森　資生堂の一連の仕事を見ていると印刷の最新技術を率先して活用されているなと思います。

青木　MG5は平グラビアを使っています。

金子　これが出たとき、春から初夏へという季節に、青年のための商品というイメージが一緒になって記憶に残っていますね。

高橋　男のラインナップがそろったという感じと、それと企業の意図が読めましたね。

青木　それ以前の男性用化粧品は、ポマードとチック。MG5でそれらにヘアトニックとかヘアリキッドだとか乳液など全部そろえて、必要品から身だしなみ商品に変えていこうとしました。宣伝費も大変なものでした。

金子　これ以前の化粧品は、使うと親父の世代に自分も入っていくという感じでしたが、MG5は自分たちの世代のためという印象が強かったですね。そして、このモノクロ。カラーテレビが普及して、カラー広告がたくさんある中でのブラック＆ホワイトは強烈でした。

青木　新聞広告のウエイトが多く、白と黒の色でとても効果的でした。

●自由商品経済への第1期生———化粧石鹸カウブランド赤箱

青木　赤箱と呼ばれるくらい大胆な赤を使っている。発売された昭和24年頃は配給制度がなくなって自由にものが買えるようになった頃ですね。

金子　戦後の自由商品経済への第1期生ですね。

青木　石鹸は匂いも機能も中身が同じだから、原色を使ったりしてデザインを変えたのでしょう。戦後、マチスなどの展覧会が開かれたことも関係あるかもしれませんね。当時からデザインは基本的には変わらずに牛の画が入っていて、牛乳石鹸ということがわかりやすいですね。今も大阪で人気が高いですね。

高橋　香りが好きで買っていて、ファンでした。

●すごいネーミング――マジックインキ

金子　ネーミングのすごさですね。

犬塚　クエスチョンマークの印象も強いですね。今もよく使っていますよ。

金子　チョーク型ハンドリングの中に筆的機能が入り込んだものといえますね。クエスチョンマークには、マジックだからというのと、この商品はいったいなんだかわかるか、という強さも感じますね。

犬塚　学校の文化祭など、とにかくお世話になりました。

青木　会社のブレインストーミングに重宝しました。これを使うと、明朝体もゴシック体もうまく書けますしね。デザイン以前に、品物が画期的でネーミングがよくてね。

金子　この名称は一般概念になってしまいましたね。

●日本の夏の風景――金鳥の渦巻

金子　世の中にモダンデザインが浸透してきた一時期、このデザインがダサイと感じることがあったけど、今また、すごいなと思います。蛍光灯の普及のときに白熱球に古さを感じ、やがてまた良さを認めたようにね。新しいものが出たとき一瞬心がそちらへ揺らぐけれど、そののち持続して復活してくるものは、やはり強いんでしょうね。

犬塚　日本の風景になってしまっていますよね。

金子　蚊帳はなくなってしまったけど、蚊取り線香は残っているね。色彩が、泥絵の具を使って描いた感じで、美術学校の時代のイメージですね。

大森　色使いが中国的ですね。

青木　ラベルの貼り箱形式だったときの名残があるデザインですね。

高橋　本当は貼り箱で、意固地にやってほしい気がします。情緒性の名残がありますね。

●工場と家庭を直結――キッコーマンと味の素の卓上容器

青木　昔は味噌なども酒屋へ行って樽から経木に盛ってもらったし、お酒も瓶のもありましたけれど、升での量り売りが多かった。樽の栓をキュッとしめると升ちょっきりになる。醬油もそうでしたね。

金子　これには、1本は買えないというのと、戦中戦後の瓶不足のためというのがありましたね。キッコーマンも明治の時代に2リットル瓶を出している。でも私の子供の頃戦時中で、一軒おいた酒屋さんで樽からアミノ酸醬油というのを買っていた思い出があります。家庭では2リットル瓶で貯え、瀬戸物屋で買った醬油差しに注ぎ変えていました。

犬塚　それが卓上瓶に変わった。キッコーマン卓上瓶の発売はいつでしたか。

金子　昭和36年です。中身入り、つまりパッケージで、しかも家庭用の本格的容器。工場と家庭を直結する工業用製品化時代の容器として、卓上瓶と名づけました。便利さが海外でも受けて、醬油の国際化にも一役果たしました。味の素の赤いキャップの瓶もこの頃すでにありましたね。これも、卓上に置いての用途化機能を進歩させた代表例ですね。

高橋　味の素の赤は、お碗の赤なのでしょうか。

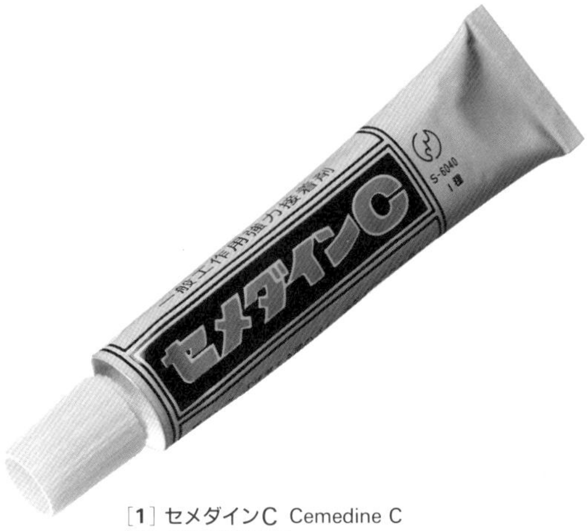

[1] セメダインC Cemedine C
[2] 1938(昭和13)
[3] 1938(昭和13)
[4] セメダイン Cemedine

[1] 特製エスビーカレー Spicy Curry Powder
[2] 1950(昭和25)
[3] 1950(昭和25)
[4] エスビー食品 S&B Foods

[1] ワンカップ大関 One Cup Ozeki
[2] 1964(昭和39)
[3] 1972(昭和47)
[4] 大関 Ozeki

[1] メリットシャンプー Merit Shampoo
[2] 1970(昭和45)
[3] 1975(昭和50)
[4] 花王 Kao

[1] キッコーマン特選卓上しょうゆ150ml びん
Kikkoman Tabletop Package
[2] 1961(昭和36)
[3] 1961(昭和36)
[4] キッコーマン Kikkoman

[1] 味の素 Ajinomoto
[2] 1909(明治42)
[3] 1950(昭和25)
[4] 味の素 Ajinomoto

犬塚　別にも赤にした理由はあるでしょうけれど、私もそれをイメージします。

金子　あの良妻賢母のマークとともに、幸せな家庭像を思わせますね。近代的小家族化は大正時代に都市部から進行し、赤缶パッケージの味の素の普及もこの時代に重なる。確か家庭という言葉が広まったのもこの頃でした。

●シャンプーのロングセラー――メリットシャンプー

青木　日本のシャンプーの17パーセントを占めたヒット商品で、ふけ・かゆみ用ということで限定した。すごいことですね。その効果があったのでしょう。品質、マーケティングも良かった。

高橋　商品コンセプトをはっきりさせたシャンプーですね。そしてファミリーユース。

金子　シャンプー、調髪液体洗剤が普及しはじめたのはいつですか。

青木　昭和31年くらいです。それまでは、フェザーシャンプーという2連のもので、男の人は1個、女の人は2個使っていましたね。それから容器入りの10円シャンプーが出ました。1回分で10円と高いものでした。

犬塚　当時ラーメンが30円ですからね。

青木　その後100円で20、30回使えるものが出ました。メリットシャンプーは、最近、デザインも色も変わりましたね。少し前の容器の独特の青緑色が印象的でしたね。

●生活に入ったモダンアート――三越の包装紙

青木　作者は猪熊弦一郎さん。大胆なパターンと大胆な色調というのは戦後のものですね。

金子　今いうところの前衛芸術、モダンアートが生活の中にずばり入ってきた象徴だと思います。しかも三越という非常に伝統的な百貨店で使われたことが面白いですね。

青木　包んでみてきれいですね。

●パッケージの魅力とは？

高橋　本当にこれらの商品すべてに時代性がありますね。

金子　記号として、シンボルとして強く心に刻み込まれているパッケージですね。単に懐かしさだけでなく、戦後このかたの暮らしにまで及んでいる。たとえ明治や戦前からあったとしても、生きてきた商品として、暮らしの中に、心の中に刻み込まれた商品の数々ですね。

犬塚　残っている商品のその理由はなんでしょうか。ここにデザインが参加しているのでしょうか。

高橋　残った理由は、受入性があったこと。でも売れるものがいいデザインかというと、営業力、量の説得力だけの話になってしまう。そうではないのでしょう。時代に淘汰されて、スタンダードになっている。原点になり、一番手になっている。商品が風景になっているものなのですよね。こうやってこれらの商品を眺めてみると、どうも人のように思えてきます。可愛い子、美人の人、あまり可愛くない子、いろいろいる。近寄りがたい美人もいるし、そんなに美人ではないけれど魅力的な人もいる。いろいろな人がいて、いろいろな世代がある、その反映という感じがする。また、これらのデザインは、ずっと変えてほしくないといいつつ、もし自分がリ・デザインをするとしたらガラッと変えてしまうだろうという気がする。多分、変えてほしくないという気持ちは、これらが自分の過去だからなのでしょうね。

［1992年10月30日収録］

Environmental deterioration is the topical issue of the world. This forces package designers to crucially review their work. In the late 50s, the traditional method of selling by pieces (candies, etc.) or by the grams (soy sauce, soy paste and oil) disappeared rapidly. As the mass production and consumption age arrived, merchandise came to be sold in pre-packaged forms. Today we are flooded with various packages, and package designers are playing an important role in our daily life.
In this book, we attempted to review how packages have come through the various changes in time, and how they have taken root in everyday living and survived to date.For this purpose, we looked back upon the history of packages in Japan by tracing back through our memories of the packages in the past.
The Japan Package Design Association sent out a questionnaire to all its members asking "What packages used in Japan after World War II are strongly imprinted in your mind?" Several hundred packages for both imported and local products were listed, the photos of part of which are included in this book.
On October 30, 1992, the five members of the Special Feature Committee had a round-table talk based on the list.

The following is the summary:

[Calpis] It expressed the people's desire for sweetness. Blue dots against the white background have a strong and refreshing impact.
[Suntory Whisky] Appreciated as a good interior item for a guest room or study room of the middle-class population.
[Cup Noodle] The vanguard of instant foods. This product was used by soldiers during the Vietnam War. Commercial advertisements were also active.
[Kirin Lager Beer] It expressed the long-lasting value of the product. The Kirin character (an imaginary giraffe) had a strong impression.
[Meiji Marble Chocolate] A chocolate candy that was launched at the beginning of the period of rapid economic growth. Children were excited by the package, which they used after eating the candy. Commercials were also vigorous.
[MG5 Hair Tonic] Cosmemtics line for men. Silver printing, and a larger cap were made available thanks to the material revolution occurring at the time. The black and white coloring was impressive.
[Cow Brand Beauty Soap (Red Box)] The first generation of the liberal merchandise economy in post-war days.
[Kincho's Mosquito Coils] It became a part of everyday life in the summer.
[Kikkoman and Ajinomoto's Tabletop Packages] Packages for seasonings gained dual functions by being used as tabletop servers.
[Merit Shampoo] Long-selling shampoo that clearly expressed the product concept.
[Mitsukoshi Wrapping Paper] It became a piece of modern art in everyday life.

What is common to these products is that they all have the strength to transcend trends, and that their packages are impressed upon the minds of consumers as the symbols of the products. They incite nostalgic feelings, however, as daily necessities they have survived competitive commercialism. They are considered to be the prototype model as well as the top-ranking products of their respective product categories. While we would like their packages to remain as they are, we feel that we would make them in completely different ways if we were asked to redesign them. Perhaps, the desire for keeping the same style might be because they are reminiscent of our past.

[1] ピース Peace
[2] 1946(昭和21)
[3] 1952(昭和27)
[4] 日本たばこ Japan Tobacco

[1] コカ・コーラ Coca-Cola
[2] 1957(昭和32)
[3] 1915(大正4)
[4] 日本コカ・コーラ Coca-Cola (Japan)

審査評

Reviews and Criticism

包装と環境――パッケージのライフ・サイクル・アセスメント

石田 修――特別審査員

地球環境保全が強く叫ばれるようになり、包装廃棄物の低減と処理が大きな問題になってきた。しかし、ここで論議されている問題は、包装廃棄物の処理方法とリサイクル方法についてのみである。パッケージは、その誕生から消滅までのライフ・サイクル、すなわち、「パッケージの揺り籠から墓場まで」をトータルで考える必要がある。これが「パッケージのライフ・サイクル・アセスメント」という考え方である。

この考え方に立つと、環境保全型パッケージには、次の諸条件が要求される。

第1に、包装材料の消費量を軽減して省資源化を促進することがあげられる。［表・1］は、1970年から1990年の間のパッケージの軽量化動向を示したものである。この表に見るごとく、パッケージの包装材料消費量は、過去20年間に大幅に低減している。

なかでも、ポリエチレン製のT-シャツ袋は、ポリエチレン製のヨーグルト・カップと高密度ポリエチレン製の2リットル繊維軟化剤用ボトルが45%も軽量化された。

第2に、パッケージの製造に要するエネルギー量を低減する省エネルギー化の促進があげられる。［表・2］および［表・3］は、1リットルおよび12オンス(355ml)包装容器の製造エネルギー消費量比較を示したものである。これらの表に見るごとく、板紙容器が最も省エネルギー性に富んでいる。これに対し、アルミニウム缶やスチール缶はエネルギー多消費型パッケージである。

そのほかに考慮すべき条件は、パッケージの保護性および安全性、充塡／シール効率、ディスプレイ性、販売訴求性、省スペース性、簡便性および再封緘性、冷蔵庫への収納適正、リサイクル性、情報表示性、収集、分別および回収性、減容率、再利用性、燃焼適正などである。

このような諸条件を考慮に入れて、パッケージのライフ・サイクル・アセスメントを実施した結果は、［表・4］［表・5］および［表・6］に示されている。この場合の評点は+5、+4、+3、+2、+1、0、-1、-2、-3、-4、-5の11段階に設定されている。

［表・1］1970-90年のパッケージの軽量化動向

パッケージ	材料	1970年［g］	1990年［g］	軽量化率［%］
ワイン・ボトル(750ml)	ガラス	450	350	−22
ビール・ボトル(250ml)	ガラス	210	130	−38
金属缶(4/4)	スチール	69	56	−19
重包装袋［1］	紙	247	215	−13
T-シャツ袋	PE	23	6.5	−70
ヨーグルト・カップ	PS	6.5	3.5	−45
繊維軟化剤用ボトル(2リットル)	HDPE	120	67	−45
パレット収縮・フィルム［2］ ストレッチ・フィルム［3］	PE	1,400	350	−75

［資料］日本包装コンサルタント㈱

［表・2］1リットル包装容器の製造エネルギー消費量比較

包装容器の種類	製造エネルギー［10^3kcal］	指数
1リットル板紙液体密封容器	137	100.0
1リットル炭酸飲料用PETボトル	1,075	784.7
1リットル3-ピース・スチール缶［4］	2,224	1,623.4

［資料］Technomic Corp.：日本包装コンサルタント㈱

［表・3］12オンス包装容器の製造エネルギー消費量比較

包装容器の種類	製造エネルギー［10^3kcal］	指数
12オンス使い捨て式ビール用ガラス瓶	496	100.0
12オンス使い捨て式炭酸飲料用ガラス瓶	605	122.0
12オンス・2-ピース・アルミニウム缶［5］(40%リサイクル)	826	166.5
12オンス・2-ピース・スチール缶	1,095	220.8
12オンス・2-ピース・アルミニウム缶詰(100%バージン)	1,377	277.6

［資料］Technomic Corp.日本包装コンサルタント㈱

注釈

［1］重包装袋
2〜3層のクラフト紙／ポリエチレンのラミネート材料から製造された大型袋で、通常、30kg以上の内容物の包装に利用される。代表的な用途は、米、麦、砂糖、食塩、化学肥料、セメントなどの包装向け。

［2］パレット収縮・フィルム
パレット(木板で構成される荷台)に製品を積み上げ、積荷が崩れないように収縮性のあるフィルムで覆いかぶせる。この時に使用されるプラスチック・フィルムがパレット収縮・フィルムで、通常、かぶせた後、加熱トンネルを通過してフィルムを熱収縮させる。
パレット収縮・フィルムは、低密度ポリエチレン(略称:LDPE)から製造される。

［3］ストレッチ・フィルム
パレットに製品を積み上げ、積荷が崩れないように伸展性のあるフィルムで覆いかぶせる。この時に使用されるプラスチック・フィルムがストレッチ・フィルムである。このフィルムは伸展性があるので、熱収縮させる必要がない。
ストレッチ・フィルムは、低密度ポリエチレン、塩化ビニル(略称:PVC)、エチレン酢酸ビニル(略称:EVA)などから製造される。

［4］1リットル・3-ピース・スチール缶
上蓋、胴体、底蓋の3部分から構成されているスチール製の金属缶。従来、スチールに錫をメッキしたブリキ板が使用されてきたが、現在は、スチールにクロームやニッケルをメッキした「錫を使わないスチール」が使用されている。胴体の接合も、以前は、半田(錫と鉛の合金)で接合されたが、現在は、電気溶接あるいはナイロン接着剤で接合されている。

［5］12オンス・2-ピース・アルミニウム缶
胴体と底蓋が一体となっており、それに上蓋の2部分から構成されているアルミニウム製の金属缶。容量は12オンス(約355ミリリットル)である。2-ピース缶には、スチール製のものもある。ビールや炭酸飲料の包装に使用されているのが2-ピース・アルミニウム缶。缶コーヒー、缶ウーロン茶、果実ジュースなどの包装に使用されているのが2-ピース・スチール缶。

［6］プラスチック・クレート
プラスチックから製造される「透し箱」で、瓶詰ビールや瓶詰清涼飲料などの1ダース入り輸送包装容器として使用される。プラスチックの種類は、高密度ポリエチレン(略称:HDPE)およびポリプロピレン(略称:PP)である。

［7］共押出多層容器
2種類以上のプラスチック樹脂を同時に押出成形して製造される容器。代表的な材料構成は、ポリプロピレン+再生材(リサイクルされたポリプロピレン)+接着剤+エチレン・ビニル・アルコール(略称:EVO)+接着剤+ポリプロピレンから成る6層構造で、ケチャップ、スープなどの包装に使用される。

［8］SiOx蒸着フィルム
酸化珪素を真空炉の中でポリエステルフィルムに蒸発定着させて製造するフィルム。アルミニウム蒸着フィルムに比較して、透明性があり、酸素および水蒸気バリア性にも優れている。味の素のレトルト食品「中華シリーズ」の包装等に使用される。

［9］段ボール・パレット
段ボールから製造されるパレット。地球環境保護の観点から、従来の木製パレットに代わって段ボール・パレットが注目されるようになってきた。しかしながら、強度の点で問題が多い。

石田 修［いしだ・おさむ］

1935年北海道生まれ。東京大学教養学部および北海道大学工学部で学ぶ。三菱化工機、サン・アルミニウム工業、テクノ・システムを経て、1971年日本包装コンサルタント㈱設立。主な著書に、「やさしいソフトウェア」(共訳)(講談社)、「POSシステムとパッケージング」(パッケージング社)、「パッケージングの知識・常識110ヶ条」(日報)、「包装用語英和・和英小辞典」(日報)など。

［表・4］評点(+2)〜(+5)のパッケージ

パッケージの種類	主要理由
段ボール容器	リサイクル適性に優れている。焼却処理が可能。ただし、リサイクル資源として、コストが問題。
折り箱	焼却処理が可能。熱エネルギーとして回収。ただし、材料資源としてのリサイクルは不可能。
貼り箱	焼却処理が可能。熱エネルギーとして回収。ただし、材料資源としてのリサイクルは不可能。
紙カップ／トレイ	焼却処理が可能。熱エネルギーとして回収。ただし、材料資源としてのリサイクルは不可能。
アルミニウム缶	リサイクル適性に優れている。物性劣化がほとんど無し。全国的なリサイクリング・システムも整備される。
プラスチック・クレート［6］	再利用が可能。廃棄後、プラスチック・パレット向け材料資源としてリサイクルが可能。
共押出多層容器［7］	リサイクル材料の再利用。省資源。焼却処理が可能。熱エネルギーとして回収。ただし、PVC、PVDC層を含まないことが前提。
共押出多層フィルム	リサイクル材料の再利用。省資源。焼却処理が可能。熱エネルギーとして回収。ただし、PVC、PVDC層を含まないことが前提。
SiOx蒸着フィルム［8］	ハイ・バリア性。焼却処理が可能。アルミ蒸着フィルムの代替製品として有望。ただし、コストが問題。
段ボール・パレット［9］	再利用が可能。リサイクル適性に優れている。焼却処理が可能。現場での組み立てが容易。
紙系緩衝材料	古紙の再利用先として最適。ただし、緩衝効果に問題あり。

［表・6］評点(−5)〜(−2)のパッケージ

パッケージの種類	主要理由
PVCボトル［21］	EC諸国一部での製造使用禁止が影響。焼却時における有害ガス発生が問題。米国での動向が注目される。
PVC加熱成形容器［22］	EC諸国一部での製造使用禁止が影響。焼却時における有害ガス発生が問題。
収縮PVCラベル	PETあるいはPO系ラベルへ転換。焼却時における有害ガス発生が問題。
木箱	木材利用効率が悪い。森林資源の保護運動家から敵視される。日本での調達が今後難しくなる。
木製パレット／スキッド［23］	木材利用効率が悪い。森林資源保護の観点から、非難が出る恐れがある。
わら縄	生態学的にはすばらしい材料だが、現実的には莫大なコストと労力を要する。

［表・5］評点(−1)〜(+1)のパッケージ

パッケージの種類	主要理由
コンポジット缶	焼却処理が可能。しかし、天地蓋が残存。ただし、日本では金属缶に比較して、コスト的に不利。
ファイバー・ドラム［10］	再利用が可能。最終的に焼却処理。ファイバー・コアー、ファイバー・チューブも同様。
パルプ・モウルド［11］	生物分解性あり。ただし、排水処理と成形精度に問題あり。現場作業性が悪く、コストも高い。
飲料カートン	回収が可能。ただし、分解、パルプ化にコストがかかる。良質の輸入板紙を使用しているので、紙資源としては最高。
バッグ・イン・ボックス［12］	外装箱が再利用可能。ただし、バッグとキャップは交換。バッグ・イン・カートンも同様。
アルミ箔容器	リサイクル適性に優れている。ただし、リサイクリング・システムの確立が急務。焼却は不可能。
金属ドラム／ペイル［13］	再利用が可能。国際的なプール・システムの確立が急務。クロージャーの改良が進展。
エアゾール	噴射剤の転換に成功。簡便性が最大の利点。容器のリサイクリング・システムの確立が急務。
ガラス容器	リサイクル適性に優れている。ただし、重量が重く、流通過程での荷役に問題あり。また、着色ガラスの再利用が難しい。
PETボトル	繊維、魚礁、人口芝などに再生可能。しかし、リサイクリング・システムの確立が急務。
HDPEボトル［14］	焼却処理が可能。熱エネルギーとして回収。将来は、モノマーを化学的回収。
LDPEボトル［15］	焼却処理が可能。熱エネルギーとして回収。将来は、モノマーを化学的回収。
PPボトル［16］	焼却処理が可能。熱エネルギーとして回収。将来は、モノマーを化学的回収。
PSP加熱成形容器［17］	回収後、減容してセメント焼成時の燃料として再利用。ただし、リサイクリング・システムの確立が急務。
OPS加熱成形容器［18］	焼却処理が可能。熱エネルギーとして回収。
HIPS加熱成形容器［19］	焼却処理が可能。熱エネルギーとして回収。
A-PET加熱成形容器［20］	PVC加熱成形容器の代替製品として有望。焼却処理が可能。熱エネルギーとして回収。
発泡プラスチック緩衝材料	回収、減容後、燃料として再利用。省資源に優れた材料。ただし、リサイクリング・システムの確立が急務。
アルミ箔複合材料	リサイクル適性がない。焼却処理時に焼却炉にアルミ屑が残存。ただし、EC諸国では、最適材料として推奨。
スチール缶	再資源化にコストがかかり過ぎる。ただし、PETラミネート缶が商業生産化されれば、問題無し。
ラミネート・チューブ	リサイクル適性がない。焼却処理時に焼却炉に屑が残存。プラスチック共押出チューブも同様。

［10］ファイバー・ドラム
クラフト紙を何十層にも巻いて円筒状にした容器。巻取り状の繊維、プラスチック・フィルム、紙、アルミ箔などの輸送包装容器として使用される。

［11］パルプ・モウルド製品
パルプ(木材繊維を砕いてつくった紙の原料)を成形してつくった製品。卵が割れないように1ダースごとに包んで輸送するのに使用されている。最近は、地球環境保護と廃棄物規制の観点から、プラスチック緩衝材料に代わってパルプ・モウルド製品が輸出包装や輸送包装の緩衝材料として使用されるようになった。

［12］バッグ・イン・ボックス
段ボール箱の中にプラスチック製の容器あるいは袋を入れた包装容器。居酒屋、レストラン、ホテルなどの業務用ワイン、日本酒などの輸送包装に使用される。

［13］金属ドラム／ペイル
金属(主として、スチール)板から製造されるドラムおよびペイル。内容量が12ガロン(約45リットル)以上をドラム、それ以下をペイルと呼ぶ。塗料、印刷インキ、石油、化学薬品、調味料などの輸送包装に使用される。

［14］HDPEボトル
高密度ポリエチレン樹脂を中空成形して製造したボトル(瓶)。主に、液体洗剤、化粧品、写真現像液などの包装に使用される。

［15］LDPEボトル
低密度ポリエチレン樹脂を中空成形して製造したボトル(瓶)。主に、マヨネーズ、トマト・ケチャップ、醤油、ソース、麺つゆ、飲料茶などの包装に使用される。

［16］PPボトル
ポリプロピレン樹脂を中空成形して製造したボトル(瓶)。主に、食用油、マヨネーズ、輸液容器、工業薬品などの包装に使用される。

［17］PSP加熱成形容器
ポリスチレン・ペーパーを加熱成形して製造した容器。主に、生鮮食肉、生鮮野菜および果物、生鮮魚肉および水産製品、冷凍食品、チルド食品などの包装にスーパーマーケットやコンビニエンスストアーで使用される。

［18］OPS加熱成形容器
延伸ポリスチレンを加熱成形して製造した容器。主に、菓子、果実、野菜、ランチ、アイスクリーム、プリン、羊羹などの包装に使用される。

［19］HIPS加熱成形容器
耐衝撃性ポリスチレンを加熱成形して製造した容器。主に、アイスクリーム、豆腐、コーヒー用ミルク、ジャム、ママレード、バター、冷凍食品、珍味および漬物などの包装に使用される。

［20］A-PET加熱成形容器
地球環境保護の観点から、焼却時に有害ガスを発生する塩化ビニルに代わって、非結晶性ポリエチレン・テレフタレート(ポリエステル)樹脂が注目されるようになってきた。しかし、現在、A-PETの価格はPVCの約1.5倍と非常に高い。

［21］PVCボトル
塩化ビニルを中空成形して製造した容器。主に、シャンプー、化粧品、液体洗剤、醤油、ソース、つゆ類、食酢などの包装に使用される。

［22］PVC加熱成形容器
塩化ビニルを加熱成形して製造した容器。主に、卵カートン、果物、珍味、漬物、佃煮などのカップやケース、医薬品および雑貨品のブリスター包装などに使用される。

［23］木製パレット／スキッド
木板から製造された荷台をパレット、当て木をスキッドと呼ぶ。資源有効利用の観点から、木製パレット／スキッドは、前述したごとく、段ボール・パレット／スキッドに代替されようとしている。

デザインの記号化、均一化志向に疑問
池田 毅

日本パッケージデザイン大賞の審査と年鑑の編集に携わって8年になろうとしている。この間に、東京を中心に日本はあらゆることで大きく変化し続けてきた。その中でパッケージデザインは、少しでも進化したのだろうか。10年以上前の協会展で展開された、地球環境や省資源を含む問題に対しての提案作品を思い出しながら、今ようやくその一部分が商品として出品されているのを見ると、日常の仕事の中で提案を続けていくのがいかに難しいことなのかを、改めて感じた。同時に企業とデザイナーの怠慢も指摘されるのではなかろうか。時代の流れと同時進行する正確な思考を、パッケージデザインの中に見つけ出すことが大切だ。

もう一点、デザインのレベルが均一化したのか、あるいは商品計画・開発の手法が同一化しているのか、それぞれの分野のパッケージデザインの顔つきが一様なのである。それは一言でいえば記号化・シンプル化した群である。差別化を計るためのシンプル志向の努力が逆に同一化して見えるのは私だけなのだろうか。パッケージデザインを構成する要素の中で情報機能が大切であることはいまさら言う必要もないが、それすらも記号化されてしまうことに不安を感じた。

記号化の方向に多勢のパッケージが向かっている中で、「フェミニテ・デュ・ボワ」が、ガラス素材の持つ特性を生かした輝きと、ある種の妖しさを表現していた。不定形の瓶は、ゆるぎないデッサン力の中で一流の造形に昇華していて見事である。感情豊かな色使いとあわせて、完成度の高い大人のデザインに仕上がっていたのが印象に強い。今年度の大賞であることに満足している。「ダスキン洗濯洗剤シリーズ」の簡潔な表現、「秋鹿」に見られる新しい日本酒の方向性も印象に残った。

ライフサイクルの長い商品に
奥村昭夫

「日本パッケージデザイン大賞1993」の審査に参加させていただき、会場に出品されている多くのパッケージを目の辺たりにして、「なんと沢山あるな」というのが正直な感想である。これまで日本ではこの会場にある何倍もの商品パッケージが市場に送り出されたことか、想像もつかない。〝マーケティングだ〟〝コンセプトだ〟〝ニーズだ〟〝ブームだ〟と多くの商品がつくられてきた時代だったと思う。それが今、経済不況でメーカーは新製品の開発を控えたり、商品をリニューアルしながらライフサイクルの長い商品にしていこうとしたりで、デザイン環境に変化が現れたと思われる。

そんな中で「日本パッケージデザイン大賞1993」に応募されたパッケージ、入選・入賞したパッケージが、1年後、2年後、市場でどれだけ生き続けられるか興味深い思いがする。できうれば、入賞したパッケージが消費者に受け入れられ、ライフサイクルの長い安定した商品に育ってほしいと思う。

社会的側面に対する賢明な提案を
桑 和美

都立産業貿易センターでの審査は、隔年ごとの審査実務の経験を生かした担当者のリードのもとに非常に円滑に進行した。今回は作品を商品ジャンル別にチェックしていったので、一次審査については、各作品の訴求ポイントについて、たとえば化粧品ならばそのカテゴリーでの時代性・造形性など、同種内の完成度と独創性を競うかたちになった。

従来の部門別方式と比較して顕著に異なる点は、実験的作品の評価が必然的に厳しくなることである。未来への提案を身上とする実験作品の中でも、市場実現化が高くまた完成度の高い作品でないと、企画の段階からもまれにもまれて製品化したパッケージと並んだ場合、どうしても見劣りしてしまう。パッケージデザインであるからには、実験作品のマーケティングおよび技術プランの具体性をもっと期待したい。

デザインの表現性という観点では、今回の応募作品はおしなべて標準以上のレベルに達していると思う。ただしパッケージ戦略的には十分成果を上げていても、「○○らしさ」に徹するあまり独創性に欠ける印象の作品には厳しい評価であった。詰め合わせタイプの贈答品群はバラエティ感覚とギフト性に共通点が多く、個々に異なる表現であるのにインパクトが弱く感じられる作品が多かった。また、家庭用品・一般雑貨・食品・医薬品類のデザインは、造形性と内容表示の強調との戦いであり、デザイン年鑑という舞台では生き残りにくいのが通例であるが、上記のカテゴリー作品の善戦が感じられた。

今回のデザイン傾向として印象深いのは、シンプルデザインということである。形状か、ロゴか、カラーか、イラストかの絞り込みがはっきりしている。嗜好品、日常品ともに限られたアイテムを効果的に演出するために、過剰なデザイン・エレメントの盛り込みを避けた表現に成功した作品が多いように思う。結果的に「こく」よりは「端正」を感じる仕上がりが、今年らしさかといえる。

また、環境対応メッセージを何らかの形で打ち出した作品の多さも、今日的といえるだろう。

パッケージが商品の顔、すなわち企業にとっては、ダイレクトなイメ

ージメーカーであることの重要性は早くから認識されていたが、今回は企業理念のメッセンジャーとしての表現が重視されるようにもなった。

特に環境対応パッケージに対する専門家を招いて、応募作品のデザイン意図を検証しながら審査に臨んだが、これは私たち審査員にとっても、大変意義のあることであった。素材の選定、形状設計など一面の知識で環境適応をうたっている作品例が、その実、廃棄・回収の現実と大きく遊離していることの半端さを指摘されることで、パッケージの社会的側面に対するよりいっそうの勉強が、質の高いデザイン設計に必要であることを痛感した。

細心の注意を払ってプロモートされ、宣伝活動と店頭というステージにデビューし、首尾よく購買カウント数を上げても、商品生命を全うした瞬間から、ゴミと呼ばれ嫌われるパッケージの一生はなかなか数奇である。産業社会にはなくてはならないモノではあるけれど、過剰・遊び・無駄の効用がもてはやされた時期の揺り戻しの現在、賢明な提案のできるパッケージデザイナー集団として、実績を積んでいきたいものである。

全ての作品の中で、最も美しかったのは、大賞となった「フェミニテ・デュ・ボワ」の非対称ボトルシェイプであった。しずくがたらりと固まった、さまざまの有機的な形は、「木の精」のコンセプトそのもので、デザイン造形の独自性と美しさに、化粧品らしさの伝統に囚われない新しさを感じた。

理屈でなく、目で視ること
高橋 敏

1,000点近くもの作品が目の前に並ぶと、圧倒されて戸惑いと同時にいい加減な視方はできないと興奮する。

審査するにあたって、まず商品としてのデザインの魅力、作者の伝えたい情緒や切り口を視る。理屈っぽく視るのは、つまらないと思う。テクニックや意図よりも、情緒や物腰。作品あるいは商品が目の前にバァンと有るべきだ。というのはここに出品されたものは、コンセプトやコスト、マーケティング、調査等の過程を経て、解いても解けそうにないパズルをやってでき上がった物たちだから、もう理屈で見るのではなく、目で視ればいい。やっぱりいいものは黙って向こうから飛び込んでくる。

今回の大賞も、もちろん大賞だという顔つきで座っていた。前回のような迷いがなく、審査する側にも良かったと思う。そして結果的に入賞作品にバラエティがあった。理論やテクニックの苦労は、でき上がった作品の前には出てきてはいけないものである。もちろん研究開発、提案をテーマにしたものは別だが。

次にプロとしての視点からの評価。これも完成された魅力のある作品であることが前提だが、たとえば和菓子や香水のパッケージの、もう何年も前からあったかのような顔つきの伝統的なデザイン、しかし今年制作されたというような、そこに隠されているよくみれば円熟した「技」にも評価をしたい。和洋を問わず伝統や趣、パッケージは生活に密着している分だけそれは大切なことだと思う。

洗練されたデザインの中での本物とは
広橋桂子

ゴミ問題、環境保護意識の高まりの中、リサイクルへの取り組みが、消費者の商品イメージ、企業イメージを大きく動かしている昨今、環境を良くすることと、デザインの調和に主眼を置いたパッケージデザインを拾い上げたいと思った。だが、包装の専門家である石田修先生のお話を伺っているうちに、省資源、省エネと生産性、経済性のかかわり合いには問題点が多過ぎて判断しかねた。

今回の審査基準は着眼性と造形性ということだった。着眼性という面では、でき上がった商品をみて、コンセプト、デザインのアイデアがどこまで徹底追求できたのか、その過程を計り識ることは困難で、一部コメントがあったものの評価しにくく、私自身の選択視は造形性に比重がかかってしまった。

入選作品のほとんどが高いレベルのデザインではあったものの、はっきり目につくような本当に革新的なものは見受けられなかった。露骨にブランドロゴが目立つものよりは、ファジー感覚というのだろうか、さりげないデザインが多い。クール、控え目、洗練の美学を具体化したものといえるだろうか。

賞候補として残った作品は、高度に洗練され、優雅、かつシンプル（シンプルとは一切の無駄を排除する厳しさ）な作品が多い。大賞を受賞した「フェミニテ・デュ・ボワ」の優美な瓶型、同系色のぼかしをあしらいながら安定感を表現している、冷静緻密な計算のもとにでき上がった作品群は、造形美という点で抜きん出ていた。特別賞の「資生堂パーラーバレンタインチョコレート」は、幾何形態に純色とロゴのダイナミックな組み合わせのデザインに新鮮さが感じられた。経営者とデザイナーとの絶大な信頼感が、このような作品を商品として世の中に送り出している事実に脱帽したい。

バブルがはじけて本物だけが残るときが来たように思われる。先人が伝統パッケージを日本の文化として残してくれたように、私達も「生活の中に文化を」という時代に生きながら、より精神性を高めるパッケージデザインを心掛けてゆきたいと思う。

●作品応募点数・入選点数

作品カテゴリー	応募点数	入選点数	賞候補点数	入賞			
				大賞	金賞	銀賞	特別賞
食品	100	35	7		1		2
菓子	135	48	6				1
アルコール飲料	89	43	11			4	1
一般飲料	101	37	14			1	1
化粧品、香水	103	48	11	1	2	2	1
化粧雑貨	88	28	4			1	
一般雑貨	33	10					
家庭用品	43	15	2			1	
医薬品	30	12	2			1	
衣料品	7	4					
電気機器、関連商品	52	26	3				2
包装紙、ショッピングバッグ、ギフトボックス	53	17	2		1		
贈答品	71	18					
輸送用ケース	12	6					
その他	16	7	1				
合計	933	354※	63※	1	4	10	8

※入選点数・賞候補点数は入賞点数を含む。

●審査期日
1992年9月28日〜30日

●作品の商品化期日
1990年9月14日〜1992年9月5日

●入選作品(入賞作品を含む)を本年鑑に収録した。

●入賞作品［作品番号 作品名］

大賞	1	フェミニテ・デュ・ボワ
金賞	2	モールド パコ
	3	セルフコンシャス シリーズ
	4	ポーラ ヴァカンシィ
	5	キッコーマン卓上しょうゆびん100ml
銀賞	6	タリアセンコーヒー
	7	UNO
	8	ポーラ システムクロノス
	9	NAN-PAO
	10	宝焼酎 スーパー純
	11	サントリー ワイン頒布会'91秋 日本の伝統美コース
	12	サントリー スキットボトル リザーブ/オールド/V.S.O.P
	13	サントリー ビア・ヌーボー 1991年夏
	14	ダンガリー
	15	ダスキン 洗濯洗剤シリーズ
特別賞	16	資生堂パーラー バレンタインチョコレート
	17	ハウス デイズコショー
	18	ゴストーコーヒー トポス
	19	Dolight
	20	秋鹿純米大吟醸一貫造り
	21	BAG・AGE
	22	カネボウ化粧品 ECシリーズ
	23	乾電池の回収/販促パッケージ

●賞候補作品

30	アルトレーゼ スパイスシリーズ
32	アルトレーゼ ハーブ入りオリーブオイル
40	ほんだし/ほんだしかつお・こんぶだし
51	カリー元年
65	資生堂パーラー フルーツゼリー/クレープショコラ/リーフパイ
68	'92バレンタイン パリコレクション
74	ヨックモック クッキーCAN夏/冬
83	シャンドーレ バレンタイン'91'92シリーズ
84	ホワイトデイ
105	サントリー ビア・ヌーボー1990
109	キリン ゴールデンビター生ビール
117	ニッカ ピュアモルト・カフェグレーン
118	サントリー ブランデーV.S.O.Pスリムボトル
125	媛の愛
138	キリン ワインクラブ ダンス
141	マキシム エステートレギュラーコーヒー
142	カフェ・デ・コロンビア クリストバル
143	プレッソ プレッソ
145	マスターズコレクション ミックススリー
148	ブレンディ 缶コーヒー
150	アサヒ コーヒーJ.O.
154	マリーム クレミュー
155	ティークオリティー
169	カルピス B&L
170	アサヒ完熟ストレート
172	アラブコーヒーパッケージ
175	キリン ポストウォーター
184	インウイ
185	アンジェリーク
187	モイストウォーター
188	ドゥ・セーズポイントメイク
204	コスメデコルテ AQトリートメントシリーズ
221	ナチュラルヘアーソープ ウィズシリーズ
234	プロセスボディーシャンプー
237	アルビオン ボディーコントロール
260	ダスキン 台所用洗剤シリーズ
269	パンシロン 胃腸薬シリーズ
300	三菱蛍光ランプ
314	ラフォーレ ショッピングバッグ
351	バンコ デザインツール

Packages and Environment
An Assessment of the Life Cycle of Packages

Osamu Ishida —— Special Judge

The needs to reduce packaging waste and to properly treat used packages (ie., disposal and recycling) are presenting serious problems as they are a cause of environmental deterioration.

Often, discussions focus on the subject of the treatment of waste packages and methods for recycling packages. The issue should be considered from the viewpoint of the life cycle of a package, from the start to the end of the life of a package. This concept is an "Assessment of the Life Cycle of Packages." Based on this concept, environmentally-friendly packages should meet the following conditions:

First, conscientious use of resource-saving methods, such as reducing the use of packaging materials. Table 1 shows the trend for reduction of package weight during the last two decades from 1970 to 1990. Through such efforts, the consumption of packaging materials has largely decreased. Most notably is the polyethelene bag for T-shirt packaging, the weight of which was reduced by 70 percent, the weight of the polystyrene cup for yogurt and the 2-liter bottle for fabric softeners made of high-density polyethelene, were lessened by 45 percent.

Secondly, energy-saving methods of manufacturing packages. Tables 2 and 3 compare the energy consumption for manufacturing 1-liter and 12-ounce (355 ml) packages by materials used. The making of paperboard containers is the most energy-saving as compared to the production of aluminum and steel cans which requires a great deal of energy.

Other factors to be taken into consideration are: secured protection of a product, filling and sealing capability, display effectiveness, appeal to consumers, space-saving capability, convenience of use, possibility of re-sealing, convenience for storing in a refrigerator, recyclability, capability of printing product information, convenience in collecting and separating waste packages, rate of reduction in volume, reusability, and burnability. The outcome of the assessment of the life cycle of packages are shown in Table 4, 5 and 6. The assessment is ranked on an 11-level scale (+5, +4, +3, +2, +1, 0, −1, −2, −3, −4, −5).

[Table 1] Reduction in Weight of Packages 1970-1990

package	material	weight (g) 1970—1990	reduction rate (%)
Wine Bottle (750 ml)	glass	450 — 350	less 22
Beer Bottle (250 ml)	glass	210 — 130	less 38
Steel Can (4/4)	steel	69 — 56	less 19
Heavy Duty Bag [1]	paper	247 — 215	less 13
T-Shirt Bag	PE	23 — 6.5	less 70
Yogurt Cup	PS	6.5 — 3.5	less 45
Fabric Softener Bottle (2 lit.)	HDPE	120 — 67	less 45
Shrink/Stretch Films for Pallets [2] [3]	PE	1,400 — 350	less 75

Source: Japan Packaging Consultants Corp.

[Table 2] Energy Consumption for Manufacturing 1-Liter Containers

package	energy consumption (10^3 Kcal)	index
1-Liter Paperboard Liquid Container	137	100
1-Liter PET Bottle for Soda Drink	1,075	784.7
1-Liter 3-Piece Steel Can [4]	2,224	1,623.4

Source: Japan Packaging Consultants Corp.

[Table 3] Energy Consumption for Manufacturing 12-Oz Containers

package	energy consumption (10^3 Kcal)	index
12-Oz Disposable Beer Glass Bottle	496	100
12-Oz Disposable Soda Glass Bottle	605	122
12-Oz 2-Piece Aluminum Can [5] (40%-Recycled)	826	166.5
12-Oz 2-Piece Steel Can	1,095	220.8
12-Oz 2-Piece Aluminum Can (100% new)	1,377	277.6

Source: Technomic Corp., Japan Packaging Consultants Corp.

Note

[1] Heavy Duty Bag
Large bags made of double- or triple-layered kraft paper, or of laminated kraft paper and polyethylene. Usually used to pack rice, wheat, sugar, salt, chemical fertilizers, and cement weighing 30 kilograms or more.

[2] Shrink Film for Pallets
Shrinkable plastic film to cover articles loaded on a pallet to prevent them from crumbling. Usually, the articles covered with the film go through a thermal tunnel in order to shrink the film. Low-density polyethylene (LDPE) is used for this film.

[3] Stretch Film for Pallets
Elastic plastic film to cover articles loaded on a pallet to prevent them from crumbling. No need for thermal shrinkage. The stretch film is made from low-density polyethylene (LDPE), polyvinyl chloride (PVC), ethylene vinyl acetate (EVA), etc.

[4] 1-liter 3-Piece Steel Can
Steel can comprising three pieces, the top, body, and bottom. Traditionally, a tin plate was employed for the top and bottom portions, but it is now replaced by such tin-free steel as chrome-plated steel or nickel-plated steel. Instead of traditional soldering (tin and copper alloy), electric welding or nylon glue are applied to join the body.

[5] 12-Oz 2-Piece Aluminum Can
12-ounce (355 ml) can consisting of the top and the integral body and bottom. Aluminum cans are used for beer and soda drinks, while steel cans are used for coffee, oolong tea and fruit juice.

[6] Plastic Crate
Crates made of high-density polyethylene (HDPE) or polypropylene (PP) used for carrying bottles.

[7] Co-Extrusion Multi-Layer Containers
Plastic container molded by extruding more than two kinds of resins simultaneously. A typical one is a 6-layer material consisting of polypropylene (PP) + recycled PP + adhesives + ethylene vinyl alcohol (EVOH) + adhesives + PP. This material is used for containers of ketch-up and soup.

[Table 4] Packages with Grades from (+ 2) to (+ 5)

Corrugated Cardboard Containers	Highly recyclable. Burnable. However, very costly as recycled resources.
Small Wooden Boxes for Food	Burnable. Retrievable as an energy source. However, not recyclable as a resource material.
Small Paper Boxes	Burnable. Retrievable as an energy source. However, not recyclable as a resource material.
Paper Cups and Trays	Burnable. Retrievable as an energy source. However, not recyclable as a resource material.
Aluminum Cans	Highly recyclable. Little deterioration in quality for reuse. Recycling systems are being organized on a nationwide scale.
Plastic Crate [6]	Reusable. Recyclable as a resource material for making plastic pallets.
Co-Extrusion Multi-Layer Containers [7]	Being processed using recycled resource materials, hence, resource-saving. Burnable and retrievable as an energy source, provided does not contain PVC and PVDC layers.
Co-Extrusion Multi-Layer Films	Being processed using recycled resource materials, hence, resource-saving. Burnable and retrievable as an energy source, provided does not contain PVC and PVDC layers.
Silicone Oxides Metallized Films [8]	Highly effective for sealing purposes. Burnable. Promising as a substitute for aluminum metallized films. However, quite costly.
Corrugated Pallets [9]	Reusable. Recyclable. Burnable. Easily assembled at the site of use.
Paper Buffer Materials	Most promising for reusing used paper. However, effectiveness as buffer questionable.

[Table 5] Packages with Grades from (− 1) to (+ 1)

Composite Cans	Burnable, however, top and bottom lids remain unburnt. Comparing to metallic cans not realistic in Japan due to high cost of manufacturing.
Fiber Drums [10]**/Cores Fiber Tubes**	Reusable. Burnable.
Molded Pulp Packages [11]	Biologically decomposable. However, problems in draining and molding accuracy. Inconvenient to assemble at the site of usage. Manufacturing costs are high.
Cartons for Drinks	Recyclable. However, processing for decomposition and reduction to pulp is quite costly. Highest quality as a resource material for paper, as the best quality of imported paperboard is used.
Bag-in-Box [12]**/ Bag-in-Carton**	The outer box (corrugated paper) can be reused by replacing the bag and the cap.
Aluminum-Foil Containers	Recyclable. Establishment of a recycling system urgently advisable. Unburnable.
Metal Drum Cans and Pails [13]	Reusable. Establishment of an international pooling system urgently advisable. Progress being made for enhanced closure.
Aerosol	Chlorofluorocarbon as a spray agent being substituted by other materials. Establishment of a recycling system for containers urgently advisable.
Glass Containers	Recyclable. However, heavy and fragile, hence, inconvenient for distribution and transportation. Colored glass hardly reusable.
Polyethylene Bottles	Reusable as fiber, man-made gathering places for fish, artificial lawn. Establishment of a recycling system for containers urgently advisable.

[8] Silicone Oxides Metallized Film
Polyester film to which evaporated silicone oxides are attached within a vacuum kiln. Compared to aluminum evaporated film, it is more transparent, and superior in blocking out oxygen and vapor. Used for "retort" packages/pouches for prepared foods.

[9] Corrugated Pallets
Pallet made of corrugated paperboard, noted as a promising substitute for wooden pallets from the standpoint of global environmental protection. However, its strength is questionable.

[10] Fiber Drum
Cylindrical container made by rolling kraft paper a few dozen times. Mainly used for carrying rolled fabric, plastic film, paper, and aluminum foil.

[11] Molded Pulp Packages
Pulp molded to pack eggs by the dozen for safe storage and transportation. Due to environmental as well as waste disposal concern, pulp molded products are increasingly used as buffer materials, in place of plastic materials, for packages for export and transport purposes.

[12] Bag-in-Box
Container made of corrugated cardboard box in which a plastic container or bag is enclosed. Used to transport liquor bottles to bars, restaurants, hotels, etc.

[13] Metal Drum Cans and Pails
Container made of metal (mainly steel) board. They are classified as drum cans when their volumes are 12 gallons (appx. 45 liters) and more, while those under this capacity are called pails. Both drum cans and pails are used for paint, printing ink, oil, chemicals, and seasoning.

[14] High-Density Polyethylene (HDPE) Bottles
Bottle made by blow molding high-density polyethylene resin. Mainly used to pack liquid detergent, cosmetics and, developing agents for photo films.

[15] Low-Density Polyethylene (LDPE) Bottles
Bottle made by blow molding low-density polyethylene resin. Mainly used to pack mayonnaise, tomato ketchup, soy sauce, Worcester sauce, noodle soup and tea drinks.

High-Density Polyethylene (HDPE) Bottles [14]	Burnable. Retrievable as an energy source. In the future, chemical retrieval of monomer will be possible.
Low-Density Polyethylene (LDPE) bottles [15]	Burnable. Retrievable as an energy source. In the future, chemical retrieval of monomer will be possible.
Polypropylene (PP) Bottles [16]	Burnable. Retrievable as an energy source. In the future, chemical retrieval of monomer will be possible.
Polystyrene Paper (PSP) Thermoformed Containers [17]	Can be used as fuel to sinter cement, after being reduced in volume. Establishment of a recycling system advisable.
Oriented Polystyrene (OPS) Thermoformed Containers [18]	Burnable. Retrievable as an energy source.
High-Impact Polystyrene (HIPS) Thermoformed Containers [19]	Burnable. Retrievable as an energy source.
Amorphous Polyethylene Therephthalate (A-PET) Thermoformed Containers [20]	Promising as a substitute for PVC thermoformed containers. Burnable and retrievable as an energy source.
Plastic Foam Buffer Material	Reusable as fuel, after being reduced in volume. Resource-saving material. Establishment of a recycling system advisable.
Aluminum Foil Composite Materials	Non recyclable. Aluminum refuse remain in the kiln when burnt. In the EC member countries, this is encouraged as the most recommendable material.
Steel Cans	Too expensive to make them reusable as a resource. However, they are reusable if PET laminate cans can be manufactured on a commercial basis.
Laminate Tubes	Non recyclable. Refuse remain in the kiln when burnt. Being substituted by plastic co-extrusion tubes.

Profile Osamu Ishida

Born in Hokkaido in 1935, Osamu Ishida studied at the Hokkaido University (Dept. of Engineering) and the Tokyo University (Dept. of Liberal Art). He has worked for Mitsubishi Kakoki Co., Ltd., Sun Aluminium Industries Co., Ltd. and Techno Systems Co., Ltd., and he established the Japan Packaging Consultants Corp. Authored Works; "Easy Software (co-translation)" (kodansha), "POS system & Packaging" (Packaging-sha), "110 knowledges & Common Senses about Packaging" (Nippo Co., Ltd.), "Japanese-English & English-Japanese Packaging Term Dictionery" (Nippo Co., Ltd.).

[Table 6] Packages with grades from (−2) to (−5)

Polyvinyl Chloride (PVC) Bottles [21]	Both production and use are prohibited in some countries of the EC. Poisonous gas emitted at the time of burning. Future developments in the United States should be watched.
Polyvinyl Chloride (PVC) Thermoformed Containers [22]	Both production and use are prohibited in some countries of the EC. Poisonous gas emitted at the time of burning.
Shrinking PVC Labels	Being substituted by PET or PO labels due to its poisonous gas emission at the time of burning.
Crate	Low lumber use efficiency. Seen with hostility from forest resource preservation activists. Foreseeable difficulty in procurement of crates in Japan.
Wooden Pallet and Skid [23]	Low lumber use efficiency. Foreseeable criticism from the standpoint of forest resource preservation.
Straw	Most recommendable material from an ecological point of view, but unrealistic as it is highly costly and laborious to collect and process.

[16] Polypropylene (PP) Bottles
Bottle made by blow molding polypropylene. Mainly used for packing edible oil, mayonnaise, intravenous medicines, and industrial medicines.

[17] Polystyrene Paper (PSP) Thermoformed Containers
Mainly used by groceries, supermarkets and convenience stores to pack fresh meat, fresh fish, fresh vegetables and fruit, processed marine products, frozen food, and chilled food.

[18] Oriented Polystyrene (OPS) Thermoformed Containers
Thermoformed container using stretchy polystyrene. Used to pack sweets, fruit, vegetables, take-out lunches, ice cream, pudding and Japanese cakes.

[19] High-Impact Polystyrene (HIPS) Thermoformed Container
Thermoformed high impact resistant polystyrene resin container. Used mainly for packages of ice cream, tofu, milk with coffee, marmalade, butter, frozen food, dainties and pickles.

[20] Amorphous Polyethylene Therephthalate (A-PET) Thermoformed Containers
Amorphous polyethylene therephthalate (polyester) is drawing greater attention as a substitute of polyvinyl chloride (PVC) that emits a poisonous gas when burnt. However, A-PET costs 1.5 times more than PVC.

[21] Polyvinyl Chloride (PVC) Bottles
Container made by thermoforming polyvinyl chloride. Mainly used for shampoo, cosmetics, liquid detergent, soy sauce, noodle soup and vinegar.

[22] Polyvinyl Chloride Thermoformed Containers
Main packages used are egg cartons, fruit cases, cups for dainties, preserved food and pickles, blister packs for medicines and sundries.

[23] Wooden Pallets and Skid
These are being replaced by corrugated cardboard pallets and skids in attempts to protect forest resources.

Designs Inclined to Signalization and Uniformity
Takeshi Ikeda

Eight years have passed since I first became a member of the jury and began participating in the compilation of the book for the Japan Package Design Award. During this period, Japan came through many changes. But I wonder if there has been any progress in package design. At an exhibition of packages organized by the Japan Package Design Association more than a decade ago, experimental packages were exhibited to provoke concern on the needs for energy-saving and environmental protection. The proposal is now being realized only partially. I have to admit that it is difficult to propose a solution to social problems through our work. Both designers and manufacturers should be blamed for their reluctance in responding to the needs of the time.
Another impression was the uniformity found in the design of many types of packages, maybe due to the standardized processes of product planning and development, or to the levelled capability of designers. They seemed to be a group of signalized and simplified designs. Simplification sought in order to be distinctive from others resulted in uniformity. I had an uneasy feeling that even the function of giving information, an important function of packages, was being signalized.
Among the many packages inclined toward signalization, the Shiseido Feminite du Bois expressed the brilliancy of glass and a sort of fascination. The irregularly shaped bottles were a beautiful sublimation of well-integrated sketches. Together with expressive hues, the package design achieved a high degree of maturity. I also had good impressions of the simple and clear expression of Duskin's detergent series, and a new direction for Japanese sake as proposed by the Akishika brand.

Products with a Long Life Cycle
Akio Okumura

My candid impression at the selection was how many they were. It is hardly imaginable how many more packages were put in the market in Japan than the ones which applied for the competition. Over the past years, a massive number of merchandise were produced under the keywords of "marketing," "product concepts," "consumer needs" and "so-and-so booms." And now that manufacturers have become less aggressive in developing new products, and are trying to prolong the life of the existing products by adding minor changes, the environment surrounding designers is also going through changes.
It will be interesting to see how many of the packages submitted to the competition for Japan Package Design Award '93, and those awarded various awards will have survived in the market a year, and a few years afterward. I sincerely hope that at least the award-winning packages will enjoy stable, long lives.
I am personally determined to do my job with care and sincerity.

Discreet Packages with Social Considerations
Kazumi Kuwa

The selection held at the Tokyo Trade Center was smoothly carried out under the leadership of a well-experienced staff. Beginning with this competition, screening was implemented according to 15 merchandise groups. Therefore, the judges compared the levels of maturity and originality of design, trendiness and formation among entries of the same merchandise category.
A remarkable change from the conventional categorization was that the evaluation on experimental works became more strict. Experimental package designs could not be compared with existing package designs which had come through rough and tough scrutiny from the beginning of the planning stages, unless they were feasible enough to be realized in the marketplace. I would expect more realistic technical planning and concrete marketing of experimental design entries.
With regard to expressive quality, all the entries were of above average quality. Therefore, even though the design quality as a package may have been good enough, those designs with little originality did not obtain high marks. The many varieties of gift assortment packages gave little impact although they employed different expressions. Designing packages for commodities, food and medicines is always a conflict between the aesthetics of form and functionality, i.e., to provide an explanation about the product. I found the entries in these categories to be quite successful in solving this dilemma.
The most impressive thing about this year's competition was that there were simple designs. Most of them focused on a particular element be it form, logo, color or illustration. This is true of both luxury items and daily commodities. Successful designs were streamlined and without excessive design elements.
Another feature was the abundance of works that expressed

the manufacturers' messages of environmental concerns. The function of packages as tools in the creation of the image of the manufacturer has been well recognized, and now, another value has been added to that, which is to be a messenger of the stance of the manufacturer.
The lecture by Mr. Osamu Ishida was very informative. His talk made us realize the fact that packages which claim to use environmentally-friendly materials and forms, in fact, apart from the collecting and storing of recylable goods, are not really using recycled items. We need to study more closely the social aspects of package design.
Packages are destined to be nuisances as soon as they have performed their roles, even if they have been designed with fullest consideration, made their contributions as a communication medium between manufacturers and consumers, and promoted the sales of the products. Today, in the backlash of a time when the effect of excessive, playful and useless decorations was lionized, we must pursue our path as designers able to create discreet packages.
The asymmetric shape of the package of Shiseido's Feminite du Bois was the most beautiful. The varied organic forms of drops aptly expressed the concept of dryads. I was deeply impressed with the originality and aesthetics of this design, which is free from the conventional themes of packages for cosmetics.

Not Reason But an Appeal to My Eyes

Ving Takahashi

Being overwhelmed at the sight of the nearly 1,000 works arranged in the screening hall, I felt embarrassed and excited at the same time, and told myself that I should be very serious in order to make fair judgements.
My first approach was to see the charm of the design as merchandise, and how the creator expressed his/her sentiment. It was not interesting to see them from a technical or conceptual point of view. Rather, aesthetic appeal was a more important criterion. All the entries were the condensation of reasonable processes such as conception, determination of cost effectiveness, marketing and research. Therefore, appeal to my eyes is the best element for judging. Naturally, good designs jumped into my eyes. The Grand Prix winning work was there as if it was predetermined to be the one. After all, the award winning works were of designs of a great variety.
The theoretical and technical considerations should not be obviously expressed in the final product, except, of course, in those works meant for R&D and proposing a future direction in package design.
Another approach was to look at the works from a professional point of view as a package designer. Mature techniques hidden behind plain looking packages had to be evaluated. These were often witnessed in traditional packages, such as those for Japanese confectionery and perfume, which appeared to have been long in existence, though they were newly designed in the year.
As package design is closely associated with the life of people everywhere in the world, I find that traditional designs and techniques are important elements to appreciate.

What is Genuine among Sophisticated Designs

Keiko Hirohashi

Amid the mounting concern regarding the issues of waste and environmental destruction, corporate approaches to resource recycling are affecting consumers' images of products as well as manufacturers. I tried to give higher marks to those packages which had good balance between environmental consideration and design quality. While listening to Mr. Osamu Ishida, an expert in packaging, however, I found it too difficult to do so as there were too many problems involved in maintaining a balance between resource and energy saving efforts and the maintenance of productivity and economic growth.
The criteria for selection this time were "conception and challenging forms." From the finished merchandise, I could hardly judge if the design concept or idea had been thoroughly realized. My eyes tended to focus more on forms. The selected works were generally of a high design standard. Yet, there were few novel ones. Nonchalant, cool, moderate, refined... these adjectives would have expressed what they were. In particular, the final candidates for the Grand Prix were highly elaborate and simple (which means to exclude verbose decorations). The elegant bottle group of Shiseido's Feminite du Bois, the Grand Prix winner, excelled for the beauty of its minutely calculated stable shape which was elegantly shaded with gradated color. The Shiseido Parlour Valentine Chocolate, a special award winner, was fresh with its logo and the combination of the three primary colors. I would like to applaud the manager and designer for their long cherished trust.
After the "bubble" economy burst, only authentic things seem to survive. Just as our ancestors gave us traditional Japanese packages, which remain to this day, we should leave behind us package designs with a high level of spirituality as evidence of the life of present day people who live in an age when "culture into daily life" has become a keyword.

Numbers of Entries and Selected Works

Category	Entries	Selected Works	Candidate for Winning Works	Winning Works			
				Grand Prix	Golden Award	Silver Award	Special Award
Food Stuff	100	35	7		1		2
Confectionery	135	48	6				1
Alcholic Beverage	89	43	11			4	1
Soft Drink	101	37	14			1	1
Cosmetics, Perfume	103	48	11	1	2	2	1
Toiletry	88	28	4			1	
Varieties	33	10					
Household Commodities	43	15	2			1	
Medicines	30	12	2			1	
Clothes	7	4					
Electric Appliances, Supplies	52	26	3				2
Wrapping Paper, Shopping Bags, Gift Boxes	53	17	2		1		
Gifts	71	18					
Shopping Containers	12	6					
Others	16	7	1				
Total	933	354*	63*	1	4	10	8

* Includes Winning Works.

- Section Period: September 28-30, 1992.
- Eligible Works: Packages commercialized between September 14, 1990 and September 5, 1992.
- This book includes all the award winning and selected works.

Scene of Judging

Winning Works [Number/Title of Work]

Grand Prix	1	Feminite du Bois
Golden Award	2	Mold Packo
	3	Selfconscious Series
	4	Pola Vacancy
	5	Kikkoman Soy Sauce
Silver Award	6	Taliesin Coffee
	7	Uno
	8	Pola Systemcronos
	9	Nan-Pao
	10	Takara-Shochu Super Jun
	11	Suntory Wine The Beauty of Japanese Tradition
	12	Suntory Skit Bottle Reserve/Old/V.S.O.P
	13	Suntory Beer Nouveau 1991 Summer
	14	Dungaree
	15	Duskin Detergent Series
Special Award	16	Shiseido Parlour Valentine Chocolate
	17	House Days Pepper
	18	Gosto Coffee Topos
	19	Dolight
	20	Akishika Junmai Daiginjo Ikkanzukuri
	21	Bag•Age
	22	Kanebo Cosmetics EC Series
	23	Dry Battery Package for Collect and Sale

Candidate for Winning Works

30	Altrecose Spice Series
32	Altrecose Olived Oil
40	Hondashi/Hondashi Katsuo•Konbudashi
51	Curry Gannen
65	Shiseido Parlour Jelly/Chocolate/Pie
68	Valentine Paris Collection
74	Yoku Moku Cookie CAN Summer/Winter
83	Chamsdorés Valentine '91 '92 Series
84	White Day
105	Suntory Beer Nouveau 1990
109	Kirin Golden Bitter Draft Beer
117	Nikka Pure Malt/Coffee Grain
118	Suntory Brandy V.S.O.P Slim Bottle
125	Hime no Ai
138	Kirin Wine Club Dance
141	Maxim Estate Regular Coffee
142	Café de Colombia Cristóbal
143	Presso Presso
145	Master's Collection Mix 3
148	Blendy Coffee
150	Asahi Coffee J.O.
154	Marim Crémieux
155	Tea Quality
169	Calpis B&L
170	Asahi Juice
172	Arab Coffee
175	Kirin Post Water
184	Inoui
185	Angélique
187	Moist Water
188	Deuxseize
204	Cosme Decorte AQ Treatment Series
221	Natural Hair Soap With Series
234	Process Body Shampoo
237	Albion Body Control
260	Duskin Dish Washing Detergent Series
269	Panshiron Medicine Series
300	Mitsubishi Fluorescent Lamp
314	Laforet Shopping Bag
351	Vanco Design Tool

あとがき

「年鑑日本のパッケージデザイン」も本書1993年版をもって5冊目となり、過去10年間にわたるパッケージデザインの秀作を記録してきたことになります。回を追うごとに作品のレベルも高くなり、当然のことながら、従来部門分けの要素となっていた販売戦略・機能・表現性などの制作意図は、それぞれを切り離して評価することがいっそう難しくなってきました。そうしたことから今回より作品分類に改定を加え、作品は受付から審査、年鑑への収録と、一貫してカテゴリー別にまとめ、作品相互の比較・評価がしやすく、より公平な判断ができるように変更されました。賞についても、総合的評価に基づく大賞、金賞、銀賞、そして作品の持つ特性に重点をおいた特別賞と、4種の賞を設定し、従来よりも受賞理由が明確になったと思います。
審査は1次・2次にわたり「着眼性(制作意図)」「造形性」の2つのキーワードをもとに投票を繰り返し進められ、賞決定に際しては熱心な討議が行われました。その詳細については、審査員による審査評をお読み下さい。
また、時代や生活の反映としてのパッケージデザインを探るための試みとして、巻末に特集を組んでみたのですが、ご覧になられてどのような印象を持たれたでしょうか。ご意見などいただければ今後の参考としたいと思います。
いうまでもなく、本書がこうして発刊されるまでには、多くの方のご尽力をいただきました。ご多忙のところ3日間にわたり審査を重ねて下さった審査員の方々、環境問題などに対して貴重なご意見をいただいた特別審査員の石田修氏、また本年鑑の主旨をご理解いただき広告掲載をしてくださった各企業に、心からお礼を申し上げます。さらに本書のエディトリアルデザインを引き受けて下さった太田徹也氏、作品撮影の阿部賢氏、特集の撮影に協力していただいたバンク・パブリシティーの皆さん、あらためてありがとうございました。最後になりましたが、六耀社の橋本周平社長はじめ編集の細川いづみさん、廣瀬雅美さんには、作品募集から審査会・作品返却も含めて、細かいご配慮をいただきました。心から感謝の意を表します。

社団法人日本パッケージデザイン協会
年鑑出版委員長 犬塚達美

Postscript

The Package Design in Japan Vol.5 is the fifth issue of a biennial book. We have recorded those package designs which were selected through the biennial contest and considered to be of excellent quality for the past decade. The standard of the designs became higher with each new issue. The elements for categorization of entry works that we set at the initial contest, i.e., sales-strategic packages, functional packages, design-oriented packages, and experimental packages, became difficult to separate. Therefore, we restructured the categorization of the contest in 1992. As a result, the application, selection, and compilation of the book are all carried out accordingly. This made comparison and evaluation of the entry works much easier and fairer. Four kinds of awards, the Grand Prix, Golden Award, Silver Award, and Special Award were created. We hope that the justification for granting these awards; the first three on comprehensive evaluation, and the Special Award for specific reasons, became much clearer than before.
The key words for selection were "design intention" and "formal beauty." Through the first screening and the final selection, voting was repeated to shortlist the works. Heated discussions were held to decide the award winners. The details of selection are described by the judges.
As one of our efforts to look for better package designs that reflect current needs, we introduced a special feature with this issue. We welcome your comments on this for our future improvement.
We are grateful to the many people involved in the contest and the publication of this book, to the judges who spared their precious three full days for selection, Mr. Osamu Ishida for helping us with his expertise in environmental issues, and advertisers to the book. Our thanks also go to Mr. Tetsuya Ota for undertaking editorial design, Mr. Ken Abe photographing the award winning and selected works and Bank Publicity Co., Ltd. for photographing the works in the Special Feature, and the editors of Rikuyo-sha Publishing, Inc. for their kind consideration and hard work in the whole process of the contest and publication of the book.

Tatsumi Inuzuka
Japan Package Design Association
Chairperson of the Publication Committee

Publication Committee of Japan Package Design Association

Executive Director	*Shuya Kaneko*
Chief Directors	*Shigeshi Omori* *Tatsumi Inuzuka*
Members	*Shigeshi Omori* *Tatsumi Inuzuka* *Shin'ichiro Satake* *Ving Takahashi* *Katsunori Sato* *Toshio Katagiri* *Nobuo Yamanaka* *Hitomi Ogawa* *Akio Maeda*

Japan Package Design Association

Head Office	CC Center, 8-15, 4-chome, Yushima, Bunkyo-ku, Tokyo Tel: 03-3815-2723•3458 Fax:03-3815-2548
Kansai Office	Heiwa Paper Co., Ltd. 4 Floor, 3-23, 2-chome, Minami Semba, Chuo-ku, Osaka Tel: 06-263-2269 Fax:06-266-9174
Chubu Office	Public Design Works Inc. 45, 4-chome, Chikaramachi, Higashi-ku, Nagoya Tel: 052-932-3856•3859 Fax:052-932-3859